FANTASTIC FOSSILS

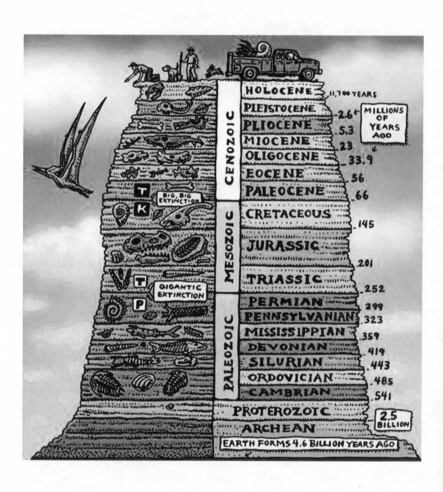

FANTASTIC FOSSILS

A GUIDE TO FINDING AND IDENTIFYING PREHISTORIC LIFE

DONALD R. PROTHERO

ILLUSTRATED BY

MARY PERSIS WILLIAMS

COLUMBIA UNIVERSITY PRESS   NEW YORK

COLUMBIA UNIVERSITY PRESS

Publishers Since 1893
New York   Chichester, West Sussex
cup.columbia.edu
Copyright © 2020 Donald R. Prothero
All rights reserved

Library of Congress Cataloging-in-Publication Data
Names: Prothero, Donald R., author. | Williams, Mary Persis, illustrator.
Title: Fantastic fossils : a guide to finding and identifying prehistoric life /
    Donald R. Prothero; illustrated by Mary Persis Williams.
Description: New York : Columbia University Press, [2020] |
    Includes bibliographical references and index.
Identifiers: LCCN 2019024030 (print) | LCCN 2019024031 (e-book) |
    ISBN 9780231195782 (cloth) | ISBN 9780231551489 (e-book)
Subjects: LCSH: Paleontology. | Fossils.
Classification: LCC QE711.3 .P76 2020 (print) | LCC QE711.3 (e-book) |
    DDC 560.75—dc23
LC record available at https://lccn.loc.gov/2019024030
LC e-book record available at https://lccn.loc.gov/2019024031

Columbia University Press books are printed on permanent and durable acid-free paper.
Printed in the United States of America

Cover design: Milenda Nan Ok Lee
Cover photo: Courtesy of Donald R. Prothero
Frontispiece: Courtesy of Ray Troll

• ◉ •

THIS BOOK IS DEDICATED TO
MY FELLOW FOSSIL COLLECTORS
AND FOSSIL LOVERS.

CONTENTS

PREFACE

Fossils are cool. Fossils are amazing. This book is about the incredible stories fossils tell us, and the staggering insights they give us into the history of the earth and the evolution of life. I hope you find this quick tour through the world of fossils both informative and enjoyable.

The idea for this book was originally suggested by my former editor at Johns Hopkins University Press, Dr. Vincent J. Burke. Working with talented illustrator Mary Persis Williams, Vince envisioned a book that would illustrate and explain fossils and paleontology to the broadest possible audience, from young adults to fossil enthusiasts of every age. But we didn't want it to be a simple "field guide" with a lot of illustrations of common fossils. There are plenty of those books, and many of them are badly out of date. Instead, I explain the fundamentals of paleontology and how to collect and identify fossils, providing practical information for any interested student or collector. Much more detailed information is presented here than you will find in picture guides to fossils. I hope this book provides enjoyable reading not just for rock hounds and hobby collectors but also for those who want to learn a bit more about fossils and how paleontologists study and understand them. It is suitable for introductory college classes in which students are learning a bit about paleontology but don't need an advanced college-level textbook. I have chosen a mix of stories and illustrations of important and commonly collected fossils along with some of the more interesting and extreme fossils that are only seen in museums.

ACKNOWLEDGMENTS

I thank Dr. Vincent J. Burke for originally suggesting this project and for recruiting Mary Persis Williams to work on it; I especially appreciated his patience as the project suffered many delays. I thank my current Columbia University Press editor Miranda Martin for her patience as the book was finally completed. I thank two anonymous reviewers for reviewing the book for scientific accuracy. I thank Kathryn Jorge at Columbia University Press and Ben Kolstad at Cenveo for their work in producing the book.

I thank my wonderful teachers in grades K–12 for giving me a sound foundation in science, especially Mrs. Helene in sixth grade, who took me on my first fossil-collecting trip to Red Rock Canyon. I also thank the great paleontologists who taught and inspired me, from my early career contacts with Dave Whistler and J. Reid Macdonald to my formal education in paleontology with Michael Woodburne, Michael Murphy, Malcolm McKenna, Gene Gaffney, and Bobb Schaeffer. Without their guidance, I would not have had the career in paleontology that I sought since I first learned about paleontology at age 4. From that age until today (over 60 years now), I never gave up, despite the difficult challenges of finding a career in this crowded profession.

Finally, I thank my amazing family: my incredible wife, Teresa LeVelle, and my sons Erik, Zachary, and especially Gabriel, who also wants to become a paleontologist someday. They put up with my long months at the computer writing this book and all the extra trips to the museums and the field to get photographs for it.

PART I

FOSSILS ARE WHERE YOU FIND THEM

Figure 1.1 ▲

Skeleton of "Black Beauty," a complete, articulated *Tyrannosaurus rex* specimen from Montana, now in the Royal Tyrrell Museum of Paleontology in Drumheller, Alberta. Paleontologist Ashley Fragomeni Hall for scale. (Courtesy of A. Hall)

FANTASTIC FOSSILS

Fossils are cool. Thanks to a huge amount of cultural exposure and the six *Jurassic Park* and *Jurassic World* movies, everyone loves dinosaurs (fig. 1.1). Lots of people enjoy collecting fossils for fun as well, whether they are fossil shells or amazing extinct creatures such as trilobites. Today you can buy all sorts of amazing fossils online and in rock shops around the world.

Fossils are cool by themselves, but they also provide a window into the prehistoric past. We can now visualize immense dinosaurs, larger than any land animal alive today; fantastic sea creatures such as the terrifying sea reptiles known as ichthyosaurs, mosasaurs, and plesiosaurs; and sea floors populated by gigantic shelled squids, huge "sea scorpions," and incredible predatory fish.

But it's not just these extinct creatures that are important. Fossils are the best clues we have to understanding the environments of the ancient past. They tell us whether a pile of rocks represents an ancient flood-plain, an ancient sea bottom, or a long-gone swamp or lagoon. We can now reconstruct ancient environments at a level unimaginable just a few decades ago. This, in turn, lets us look at issues like past climate changes: episodes of warming, when greenhouse gases warmed the planet and oceans drowned the continents, and times when the world was frozen from the poles almost to the equator. We can reconstruct ancient ocean currents and ancient weather patterns as well. Finally, fossils are the primary means used by geologists to date rocks. Without fossils we could not tell geologic time.

Fossils are really important for our civilization for other reasons too. Not only do they tell us about ancient climates, but fossils (and the life they represent) *controlled* our climate in the geologic past. Ancient bacteria were responsible for giving our planet the oxygen we breathe—and algae in the oceans remain the major providers of oxygen. Without these organisms, this planet would have been as lifeless as Mars or any other body in space. The next time you see some pond scum, thank them for your ability to breathe!

Thick deposits of coal (fig. 1.2A) are the remnants of huge swamps that did not decay; as the trees fell, their material was locked up in the earth's crust and became the black rock we burn today. Ancient limestones (fig. 1.2B) are remnants of tropical lagoons similar to those in the Bahamas or the South Pacific today. Their trillions of shells also locked up a huge amount of carbon in the earth's crust as calcium carbonate, or calcite. Together, coal and limestone are the great regulators of carbon dioxide in our atmosphere. When they lock up lots of carbon in the crust, the planet can go into an icehouse state. When we burn lots of coal (as we do today), the trapped greenhouse gases are released into the atmosphere and global warming occurs.

Contrary to the popular myth, all the oil and gas that we use did not come from decayed dinosaurs. Instead, it is from the shells of trillions of plankton that once lived on the surface of the ocean and are now buried in deep-sea muds. Just like burning coal, when we burn these fossil fuels, we release the trapped carbon dioxide that was entombed in the crust, which contributes to climate change.

So fossils are not just cool; they tell us amazing stories about the past. The life that fossils represent helps us know why our climate has changed

Figure 1.2 ▶

Many extinct organisms are responsible for controlling our climate: (*A*) A coal seam, which is made of the remnants of thousands of plants that turned to stone instead of decaying. (*B*) A fossiliferous limestone from the Late Silurian Wenlock Limestone, Dudley, UK. The large triangular object (*right center*) is the tail segment of the trilobite *Dalmanites*. The honey-combed objects are pieces of the coral *Favosites*. The long cone (*left center*) is a straight-shelled orthocone nautiloid. There are numerous brachiopods such as *Atrypa*, *Leptaena*, and rhynchonellids, as well as branching bryozoans such as *Favositella*, and stems of crinoids. ([*A*] Photograph by the author; [*B*] courtesy of Wikimedia Commons)

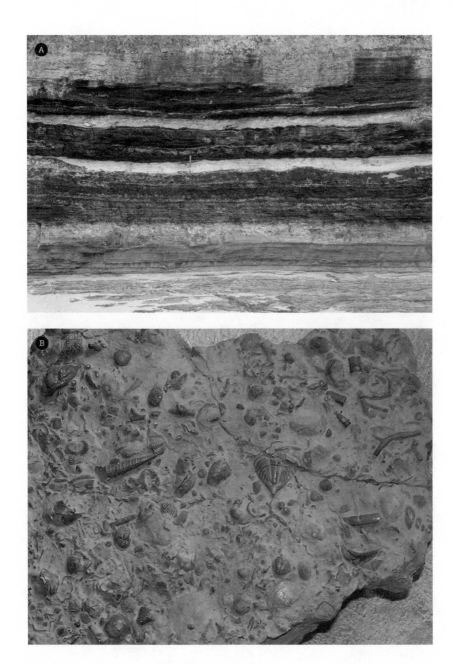

and the vital role oxygen has played in life on this earth. As we burn those same fossils, which are now extracted as oil and coal, we are changing our climate yet again. And we may be transforming the earth in ways that might not be good for our future.

HOW ARE FOSSILS FORMED?

What is a fossil? A fossil is the preserved remains of any once-living organism. Some fossils, such as the freeze-dried mammoths and other Ice Age mammals found in the Siberian permafrost, are almost completely preserved, and their original tissues are largely intact (fig. 2.1A). In some cases, the meat was so well preserved that Russian sled dogs and even humans could eat it when these mammoths began to thaw. An Ice Age woolly rhinoceros was found completely pickled in tar in Starunia, Poland, and most of its original tissues were preserved (fig. 2.1B). Scientists have tried to extract the DNA from these exceptionally well-preserved specimens, but their success has been limited because DNA degrades rapidly after death. Even when the animal is well preserved, contamination from bacterial DNA is present in the carcass. Despite this degradation, we have learned a lot from these specimens. They reveal the color, fur patterns, and even the stomach contents for these long-extinct beasts. Apparently, woolly mammoths had a fondness for buttercups.

Slightly less well-preserved are specimens from places like the Rancho La Brea tar pits (and similar tar seeps in other places in California, Peru, and around the world). Although the flesh is gone, the original bone material is present, but it has been pickled in tar (fig. 2.1C). Early attempts to discover the DNA sequences for saber-toothed cats and other Ice Age mammals from these specimens were false alarms; most of the early published studies were found to be based on bacterial DNA that contaminated the bones after the animal had decayed. The original shell material of many molluscs

Figure 2.1 ▲

There are many different kinds of fossilization. (*A*) In Siberia and Alaska, Ice Age mammals were freeze-dried and mummified, like this baby woolly mammoth. (*B*) In certain tar seeps, complete animals may be pickled in tar like this complete woolly rhinoceros from Starunia, Poland. (*C*) In the famous La Brea tar pits in Los Angeles, the bones are pickled in tar. (*D*; color figure 1) These Cretaceous ammonites still have their rainbow-colored iridescence from "mother of pearl" aragonite preserved in their shells. (*E*) Insects and other animals are occasionally trapped in tree sap, which can harden into a rock called amber, allowing extraordinary preservation. (*F*) The original woody texture of this piece of wood from the Petrified Forest in Arizona has been replaced by minerals such as silica (silicon dioxide) and some red iron oxides ("rust"). ([*A, C–D, F*]) Photographs by the author; [*B, E*] courtesy of Wikimedia Commons)

Figure 2.1 ▲
(continued)

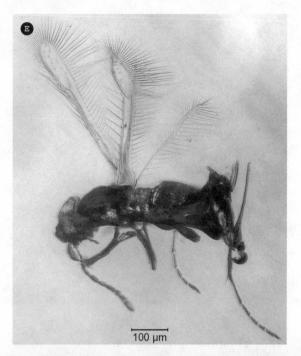

100 µm

Figure 2.1 ▲

(*continued*)

(especially Ice Age molluscs) has survived intact, including the organic material. Other examples of excellent preservation are shells with their original "mother of pearl" layer (the form of calcium carbonate, $CaCO_3$, known as aragonite), as aragonite is unstable and usually changes to a more common form of calcium carbonate, known as calcite (fig. 2.1D).

Insects and other creatures trapped in amber, including lizards and frogs, also provide famous cases of extraordinary preservation (fig. 2.1E). They are beautiful in their detail, with even the most delicate wings and bristles and hairs intact. Nevertheless, their organic material has completely degraded into just a carbon film. Contrary to the premise of the *Jurassic Park/ World* books and movies, no original DNA is preserved in amber fossils. Not only is it impossible for dinosaur blood to survive in the gut of Mesozoic mosquitoes encased in amber, but even the original insect remains have no original DNA because amber is quite porous to chemicals from the outside.

Another common form of fossilization is *permineralization*. Many biological tissues (bones, wood, and so on) are full of pores and canals where a chemical can percolate through and crystallize within the fossil, preserving the original structure. This kind of preservation is common in petrified wood (fig. 2.1F), which can be completely permeated by carbonate or silica, in which the original growth rings and even cell structure can be seen.

Another common pattern in fossilization is *recrystallization*. Many shells are made of unstable minerals, such as aragonite ("mother of pearl"), and those minerals revert to a more stable form over time and with burial. For example, aragonite recrystallizes into calcite. Fine-grained granular calcite can recrystallize into large, coarsely crystalline calcite. The original shape and appearance of the fossil remains the same, but microscopic examination shows that the crystal texture and size has changed from its original form.

The most common form of fossilization is *dissolution and replacement*. Most fossil shells do not contain the original shell material. Instead, that material dissolved away, leaving a void that was replaced by new minerals (fig. 2.2). In addition, the internal cavity within the fossil can fill with crystals of calcite, silica, or even sand grains, which can form an *internal mold* (also known by the German word *steinkern* [stone cast]) that preserves the details of the shape and texture of the internal part of the shell, or even the brain cavity within a skull. The mineral in the original fossil is usually

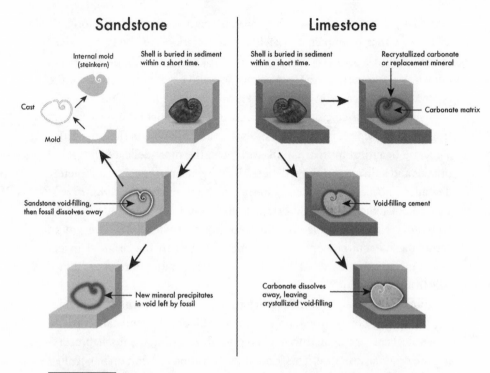

Sandstone

Internal mold (steinkern)

Shell is buried in sediment within a short time.

Cast

Mold

Sandstone void-filling, then fossil dissolves away

New mineral precipitates in void left by fossil

Limestone

Shell is buried in sediment within a short time.

Recrystallized carbonate or replacement mineral

Carbonate matrix

Void-filling cement

Carbonate dissolves away, leaving crystallized void-filling

Figure 2.2 ▲

The stages of dissolution and replacement of a fossil in a sandstone or limestone matrix. The original shell can be dissolved and replaced with a different chemical. In addition, the internal cavity of the fossil can be filled with sand or a cement made of crystals of silica or calcite. That void-filling material can weather out as an internal mold of the fossil's internal cavity. (Illustration by Mary Persis Williams)

replaced by the same mineral (especially calcite). But a shell originally made of calcite may also have been replaced by silica (SiO_2, or silicon dioxide), the mineral dolomite [$(Ca, Mg)(CO_3)_2$], or even sparkly crystals of "fool's gold" (fig. 2.3A), which is the mineral pyrite (FeS_2). We know that no living creature uses pyrite in its shell, so this is clearly a later replacement (fig. 2.3B).

Finally, certain kinds of fossils are preserved as flattened pancakes, with the original film of their soft tissues still visible in the process of *carbonization*. This type of preservation is particularly common in deep-water black shales or in swampy settings, where fossils were buried in stagnant, low-oxygen conditions and then covered in mud, which prevented

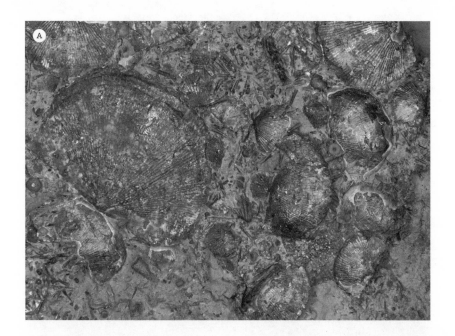

The original calcite mineral in shells may be replaced by some other nonbiological mineral, such as pyrite, or iron sulfide (fool's gold): (A) brachiopods replaced by metallic pyrite; (B; color figure 2) an ammonite shell completely replaced by pyrite. (Courtesy of Wikimedia Commons)

Figure 2.4 ▲

Many fossils are preserved as they were crushed in shales, leaving only the carbonized film remains of their original tissues. (*A*) Fossil ferns and other plants preserved as a carbon film in sandstone. (*B*) Complete articulated skeleton of the marine reptile known as the ichthyosaur, with the body outline preserved as a carbonized film around it. (*C*) Fossil bird from the Eocene Green River Shale showing feathers and other soft tissues. (Courtesy of Wikimedia Commons)

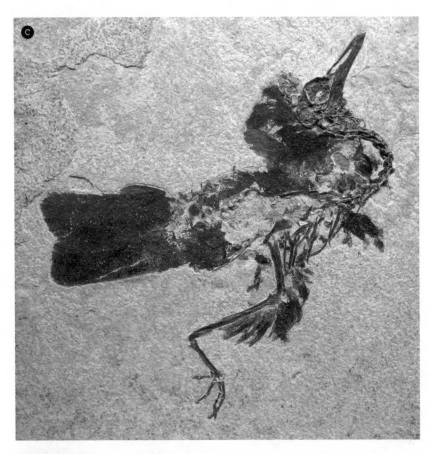

Figure 2.4 ▲
(*continued*)

scavengers from eating them. Most plant fossils preserved in coal beds are reduced to carbonized films (fig. 2.4A). And the body outlines of creatures such as ichthyosaurs (fig. 2.4B) and fish, as well as the soft tissues of creatures like trilobites and graptolites, are still preserved in many famous deepwater shales. Stagnant lake shales, such as the Eocene Green River Shale of Wyoming, Utah, and Colorado (fig. 2.4C), or the Florissant Fossil Beds in Colorado, also preserve complete insects and the leaves of plants this way.

03

WHAT KINDS OF ROCKS YIELD FOSSILS?

Many kids are fascinated with dinosaurs and other prehistoric creatures. Some even dig holes in their backyards or driveways looking for dinosaur bones. Most eventually become discouraged and give up because dinosaur bones are extremely rare, and they are found in only a few places on Earth.

If you wanted to find fossils, where would you look? Why are certain rocks and certain places good for finding fossils, whereas others have none at all?

There are three types of rocks—sedimentary, metamorphic, and igneous rocks—but fossils are primarily found only in *sedimentary rocks*. These rocks are made from the loose grains of sand, gravel, mud, or other particles that weather out of the hard bedrock and are deposited in rivers, floodplains, or at the bottom of the ocean. When animals and plants die, their hard parts (bones, shells, wood) may be buried. If the conditions are right, these hard parts will be deeply buried and then be covered by loose sediments (fig. 3.1). Over time the sands are cemented together by minerals in the groundwater to become sandstone (fig. 3.2A), or the soft mud grains are squeezed and compressed until they become a hard splintery rock called shale (fig. 3.2B).

These sedimentary rocks might then be deeply buried in the earth's crust. Millions of years later, these ancient rocks might be uplifted to the surface and crumpled upward by the collision of continents to form a mountain belt. Or they might be tilted on their side, and erosion will expose the ancient sediment. Or they might be brought to the surface by erosion, and rain and frost and wind will break down the fossils as well as the rock

The dinosaur lives out its life before eventually dying in a muddy riverbank.

Time

The body is exposed to scavengers and the elements until bleached bones are all that remain.

As time passes, the water level rises and buries the skeleton in sediment.

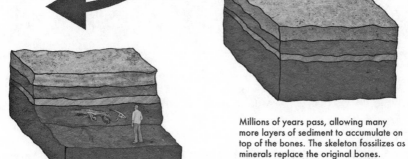

Millions of years pass, allowing many more layers of sediment to accumulate on top of the bones. The skeleton fossilizes as minerals replace the original bones.

Erosion exposes the layer of rock hiding the skeleton. Lucky fossil hunters may now dig it out and begin their studies of the dinosaur.

Figure 3.1 ▲

How a dinosaur skeleton becomes a fossil. (Illustration by Mary Persis Williams)

Figure 3.2 ▲

(A) Mollusc fossils inside a sandstone. (B) A dense mass of mollusc shells, including razor clams and many other kinds of molluscs, from the middle Miocene Olcese Sand, Bakersfield, California. ([A] Courtesy of Wikimedia Commons; [B] photograph by the author)

surrounding them (fig. 3.3). This has been happening for millions of years, and most of the fossils that were once buried have already been exposed by erosion and were destroyed before humans got around to collecting them. Only in the past 200 years have humans (especially paleontologists) been

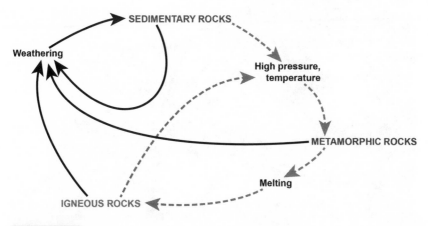

Figure 3.3 ▲

Rocks are formed by various natural processes. Of the three types of rocks, fossils are generally found only in sedimentary rocks. (Illustration by Mary Persis Williams).

actively looking for, collecting, and preserving fossils before they are lost forever. Fossil collectors have visited only small parts of the earth more than a few times. Even today, large areas of unexplored land remain in remote regions, and most fossils there will be lost before any human sees them in time to save them.

In addition to sands and gravels and muds, another common kind of sedimentary rock is limestone, and it is literally made of fossils—mostly the broken fragments of shells of sea creatures that lived millions of years ago (see fig. 1.2B). If you happen to be collecting in an area where limestones are common in the bedrock, fossils are everywhere. However, most of these fossils may be highly fragmentary and are not worth collecting.

Igneous rocks are formed by the cooling of magma, or molten rock, that comes up from the hot deep interior of the earth (fig. 3.4A). This can happen when a volcano explodes and scatters volcanic ash across the landscape (as happened at Mount St. Helens in Washington in 1980), or when lava flows pour out of an erupting volcano (as happens on Kilauea on the Big Island of Hawaii nearly every year or so). The magma might remain underground without ever erupting from a volcano, but instead cool in a deep magma chamber until it is a hard crystalline rock like granite (fig. 3.4B). Either way, igneous rocks rarely preserve fossils. Hot magma usually incinerates or vaporizes the soft tissues of an animal or plant without leaving a trace.

Figure 3.4 ▲ ▶

(*A*) Igneous rocks include lava flows, often with a pattern of polygonal fractures called columnar jointing, which are formed as they cool from a molten state. These lava flows are from the Isle of Staffa in Scotland and have been carved by the waves to create Fingal's Cave. (*B*) Magma chambers produce coarse-grained igneous rocks, such as this granitic boulder from the Sierra Nevada Mountains in California. (*C*) Metamorphic rocks have been under intense pressure and hot temperatures, which results in a layered structure known as foliation. This 1.7 billion-year-old Mendenhall Gneiss from the San Gabriel Mountains is foliated but also has been intensely folded. (Photographs by the author)

Figure 3.4 ▲
(*continued*)

In only a few cases has volcanic ash blown for a long distance and buried a creature, preserving it in some way.

The third class of rocks are *metamorphic rocks*. They are formed when igneous or sedimentary rocks descend deep into the earth's crust and are put under immense pressures and extremely hot temperatures (fig. 3.4C). These conditions transform the original rock into a new rock with new minerals and a new fabric. Any remains of plants or animals are destroyed in this process, so there are no fossils to be found in metamorphic rocks (unless the rocks are just barely metamorphosed).

04

WHERE DO YOU FIND FOSSILS?

Many experienced fossil collectors begin by looking at a geologic map, which shows the name and the age of the rock units exposed on the earth's surface. If the rocks are metamorphic or igneous, the paleontologist ignores them. They focus instead on sedimentary rocks, especially those in mountainous areas where they may have been uplifted and eroded.

The next thing you must determine is whether the exposed sedimentary rocks are the right age to contain fossils. Again, a paleontologist begins by looking at a geologic map that not only shows the location of sedimentary rocks but also the age of the rock formations (fig. 4.1). If you are looking for dinosaurs, for example, you would only look in rocks from the Mesozoic Era, or the "Age of Dinosaurs" (250–66 million years ago). Most dinosaurs come from rocks of the Jurassic Period (140–200 million years ago) or the Cretaceous Period (66–140 million years ago). This greatly limits the number of places you can look because only a small percentage of the exposed rock on the earth's surface are Jurassic or Cretaceous in age.

In addition, the geologic map may also indicate what sedimentary environment formed the rocks. Dinosaurs are often found prehistoric river floodplains or possibly in ancient lakes, where they lived and also died and were buried (see fig. 3.1). You would not expect to find good dinosaur fossils in rocks that represented the open ocean because all dinosaurs were land creatures. (Marine reptiles such as ichthyosaurs, plesiosaurs, and mosasaurs lived in the seas of the Mesozoic, but they are not dinosaurs, even if the public thinks so.) This quickly limits your options even further because

Figure 4.1 ▲

A geologic map of an area shows what kind of bedrock (shown by the bands of different shades) is found beneath the plants and loose soils and sediments. Certain key fossils (as shown in the photo) are typical of most fossiliferous sedimentary formations, and the occurrence of these index fossils allows a paleontologist to identify the formation, correlate it with other formations, and determine its age. (Photograph by the author)

most of the good exposures of Jurassic and Cretaceous rocks are marine rocks, which rarely produce dinosaur fossils. Only a handful of places on Earth have exposures of nonmarine Jurassic and Cretaceous rocks that produce dinosaur fossils, and they are in desert badlands.

A good example of how paleontologists select a site was described by my friend, classmate, and coauthor Neil Shubin in his best-selling book, *Your Inner Fish: A Journey into the 3.5-Billion-Year History of the Human Body*. He and his colleagues were looking for a fossil that was intermediate in age between the earliest amphibians that had been collected and the most advanced youngest amphibian fossils known. They looked on geologic maps for rocks of a certain time span (385–365 million years ago) in the Late Devonian Period, which already had good fossils of very primitive amphibian-like fish from 385 million years ago as well as fossils of more advanced fish from 365 million years ago. After scouring the geologic maps of the world, they found only three areas of the right age, and the right sedimentary environment (shallow marine sandstones and shales, with some from rivers or deltas). Two of them (in Pennsylvania, and in Spitsbergen and Greenland in the Arctic) had already been explored and collected. But the third place, in the Canadian Arctic, had never been studied. They raised money for a quick visit to the area, and they found bone scraps. To mount a full-scale Arctic expedition, they had to raise millions of dollars more, which they did for several summers in a row. Only after the third year of very hard, very expensive, very dangerous work dealing with harsh weather and marauding polar bears did they find the fossils of *Tiktaalik*, the famous transitional fossil between fish and amphibians (fig. 4.2). After this bonanza, they went back several more times, finding many more specimens of *Tiktaalik* along with a number of other fish and animals that lived in this ancient river delta 375 million years ago.

If all this seems like very hard work, it is. To discover rare fossils like dinosaurs or *Tiktaalik*, you can spend months in the field looking and spend a lot of money and you may still find nothing at all. Most paleontologists use a much more reliable method—they go to places where the fossils they are looking for have already been found. The best way to make sure you find something is not to gamble on unknown areas on a geologic map but to go to known fossil beds (many websites list these fossil beds) and see what you can find.

The transitional fossil between fish and amphibians is known as *Tiktaalik*. Neil Shubin and the crew found it in the Canadian Arctic after consulting geologic maps to see where rocks of the right age for this transition might be preserved. (Courtesy of Wikimedia Commons)

In the western United States, many localities are known to produce lots of fossils. For example, the Big Badlands of South Dakota have been producing fossil mammals and turtles every year since 1848. And the Miocene rocks of western Nebraska are legendary for their richness and long sequence of Oligocene and Miocene fossil mammals. For example, dense bone beds mostly filled with the bones of the small rhinoceros *Menoceras* (fig. 4.3A) can be found at Agate Springs Fossil Beds National Monument in Nebraska.

Discoveries also occur when the paleontologist is hard working and well prepared. Mike Voorhies of the University of Nebraska State Museum was prospecting for fossils among the riverbank outcrops in northeastern Nebraska one day in 1977 when he found a bit of rhino skull poking out of the rock. He dug in further and found the skull of a baby rhinoceros. As he kept digging, he found its complete skeleton, then he found the skeleton of its mother next to it. Within a few months, he and his crew had found

a complete bone bed of hippo-like rhinos know as *Teleoceras* that had died in volcanic ash, which suffocated them and buried them in life poses (fig. 4.3B). They were so well preserved that even the tiniest bones in the throat were still there, as well as the fossilized seeds of their last meals. Some of the female rhinos had fetuses still in them, and others had baby rhinos in nursing position when they died. In addition to rhinos, this locality produces a few horses, extinct musk deer, and even cranes. The locality was nicknamed "Rhino Pompeii," but today it is called Ashfall Fossil Bed State Monument. You can visit the bone bed as it is still being excavated, protected from the weather and vandals by a huge building known as the "Rhino Barn."

Most of the rich dinosaur-bearing beds of the world are well known and have been collected off and on for years. For example, in 1909 paleontologists Earl Douglass and William J. Holland of the Carnegie Museum of Natural History in Pittsburgh were prospecting outcrops of the Morrison Formation, a well-known Upper Jurassic dinosaur-bearing unit in the Uinta Basin of northeastern Utah. Holland described their work this way:

We decided that we would set forth early the next day with our teams of mules and visit the foot-hills, where Hayden had indicated the presence of Jurassic exposures. We started shortly after dawn and spent a long day on the cactus-covered ridge of Dean Man's Bench, in making our way through the gullies and ravines to the north. . . . The next day we went forward through the broken foot-hills which lie east and south of the great gorge through which the Green River emerges from the Uinta Mountains on its course to the Grand Canyon of Arizona. As we slowly made our way through the stunted groves of pine we realized that we were upon Jurassic beds. We tethered our mules in the forest. Douglass went to the right and I to the left, scrambling up and down through the gullies in search of Jurassic fossils, with the understanding that, if he found anything he was to discharge the shotgun which he carried, and if I found anything, I would fire the rifle, which I carried. His shotgun was presently heard and after a somewhat toilsome walk in the direction of the sound I heard him shout. I came up to him standing beside the weathered-out femur of a *Diplodocus* lying in the bottom of a very narrow ravine in which it was difficult to descend. Whence this perfectly preserved bone had fallen, from what stratum of the many above us it had been washed, we failed to ascertain. But there it was, as clean and perfect as if it had been worked out from the matrix in the laboratory. It was too heavy for us to shoulder and carry away, and possibly even too heavy for the light-wheeled vehicle in which we were traveling. So we left it there, proof positive that in that general region search for dinosaurian remains would probably be successful.

Holland's prediction came true a year later, on August 17, 1909, when Douglass was working in the same area with a local Mormon farmer, George "Dad" Goodrich. Douglass climbed the ridge above where the femur had been found the year before and looked down. "At last, in the top of the ledge where the softer overlying beds form a divide—a kind of saddle—I saw eight of the tail bones of *Brontosaurus* in exact position. It was a beautiful sight." He and Goodrich went back to town and recruited more helpers, then they began to quarry out the bones. "I have discovered a huge dinosaur *Brontosaurus* and if the skeleton is as perfect as the portions we have exposed, the task of excavating will be enormous and will cost a lot of money, but the rock is that kind to get perfect bones from." This message brought Holland to Utah, and when he saw what Douglass had, he immediately telegraphed Andrew Carnegie, the rich man who had founded their museum, to get him to agree to fund the excavation.

The nearest town was tiny Vernal, Utah, over 20 miles (32 km) away, so for the next 13 years (1909–1922), Douglass took up permanent residence in the area near Carnegie Quarry. There he and his crews lived and worked year round except when the weather was unbearable. He even brought his young wife and one-year-old baby out to live with him, first in a small tent heated by an iron stove, but eventually they lived in a homesteader's log cabin with a garden and cow and chickens and everything the family needed. First they exposed the "Brontosaurus" he originally discovered, only to find the neck twisted back into the rock—and the skull missing. Nevertheless, it was nearly complete and about 98 feet (30 meters) long, with a tail over 30 feet (9 meters) in length. When it was shipped to Pittsburgh, cleaned, and mounted, Holland described it and named it *Apatosaurus louisae* in honor of Carnegie's wife Louise.

Once the first skeleton had been removed, Douglass and his men found three other sauropod skeletons nearby. They realized that the bone bed was a thick ledge of sandstone that was tilted almost vertically. They blasted away the overburden of soft Morrison shales and trenched down to expose the top surface of the tilted sandstone layer. Eventually the trench was

600 feet (180 meters) long and 80 feet (24 meters) deep. Between 1909 and 1922, Douglass and his men removed the top half of the sandstone wall, which was more than 300 feet (90 meters) long and 75 feet (23 meters) high. They also excavated the east and west side of the huge wall of sandstone.

They recovered 700,000 pounds (315 tonnes) of fossils and took them by buckboard wagon to the nearest rail stop in Dragon, Utah, over 50 miles (80 kilometers) away. Altogether they found more than 20 skeletons and fossils representing about 300 additional individual dinosaurs. Apparently, the sandstone layer had once been a sand bar in a Jurassic river, and portions of carcasses had floated down to that spot and then become buried. In 1922, William Holland retired, and Andrew Carnegie died, so the funding dried up. The Carnegie Museum was crammed to the limit, with more than 300 tons (270 tonnes) of bones that had not yet been prepared or cleaned, so the museum decided to end the excavation and close the quarry. Douglass then began working for the University of Utah and collected another 33 crates of specimens, including a complete *Allosaurus*. However, they never offered him a position, and he died in poverty in 1931 without seeing his dream come true.

Douglass's dream was to see the Carnegie Quarry made into a national monument. They had removed only half of the original wall of sandstone, and he knew that the locality had amazing potential. Douglass wanted the rest to be left in place as a permanent monument for people to see dinosaur bones as they are found in the field. As he wrote, "I hope that the Government, for the benefit of science and the people, will uncover a large area, leave the bones and skeletons in relief and house them in. It would make one of the most astounding and instructive sights imaginable." Douglass tried to buy the mineral rights to protect the site, but the courts rules that dinosaur bones were not minerals. But Holland had a powerful friend, Charles Doolittle Walcott, a paleontologist who was also head of the Smithsonian. Walcott convinced President Woodrow Wilson to designate the quarry area as Dinosaur National Monument in 1915.

The monument was isolated in the middle of the wilderness of Utah and remained primitive and undeveloped for years because there was almost no way to reach it in the days before cars were common and roads paved. During the Depression, crews of unemployed men came from the WPA to deepen and enlarge the quarry. Nothing much was built on the site during the Second World War, but in the 1950s the Park Service surveyed the area

Figure 4.4 ▲

Dinosaur National Monument Quarry Visitors Center, showing the wall of dinosaur bones left where they were found but excavated in relief. (Photograph by the author)

and determined it was worth developing. A modern glass-sided building was finished in 1958, with its north wall made of the sandstone layer full of dinosaur bones.

Over the years, the building has become one of the most popular national monuments in the country (fig. 4.4). About 400,000 visitors a year come to northeastern Utah to gawk at the wall of dinosaur bones that are slowly being excavated in relief and left in place (unless they need to be removed to expose more bones). In 2009, the quarry building was closed because it was beginning to sink into the soft Morrison shales below its foundation, and that was tearing it apart. A much larger Visitor's Center was built down on the flats below the quarry (near where Douglass had his log cabin), and a shuttle service now brings up the flood of visitors coming to the tiny parking lot below. The quarry building was rebuilt with 70-foot steel pilings driven deep into the harder bedrock, and it reopened in 2011. It only houses the paleontological exhibits. The main Visitor's Center down on the flats has exhibits about the rest of the monument, plus a gift shop, offices, and storage for the specimens removed from the quarry.

The rocks are constantly being eroded and exposed each year when the rains and snows come and go, so some new fossils might be exposed the following spring and summer. The paleontologists who find fossil dinosaurs usually work in a particular area year after year, and sooner or later they begin to find something worth the huge amount of time and expense that has been invested in this difficult work.

05

DATING FOSSILS

How old is your fossil? As we saw in chapter 4, this question is fundamental not only to identifying your fossil but also to knowing where to look for fossils. If you're looking in beds of the wrong age, you won't find the right kinds of fossils—or maybe no fossils at all.

Geologists and paleontologists determine the age of rocks and geologic events using two methods: relative dating and numerical dating. The first method, *relative dating* or relative age, tells us that geologic event A is younger or older than geologic event B. Four fundamental principles are used to help us determine the relative age of an event (fig. 5.1). The primary way geologists do this is by using the *principle of superposition* (fig. 5.1A), which was first proposed by the Danish scholar Nicholas Steno in 1669. In any layered sequence of rocks (usually layered sedimentary rocks, although it applies to layered lava flows as well), the oldest rocks are at the bottom of the stack, and each layer above it is progressively younger. In other words, the stack of rocks goes from older at the bottom to younger at the top. You can't stack one layer on top of another if the lower layer isn't already there first. A good analogy is the stack of papers on a messy desk or table. If they just keep accumulating through time without being turned over or sorted out, then the oldest papers will be at the bottom of the stack, and the most recent ones will be at the top. Thus, if you are looking at the impressive pile of layers in the Grand Canyon, the oldest ones are always at the bottom, and each layer above it is younger. They are like the pages in a book, with the first page at the bottom of the stack, and the last at the top.

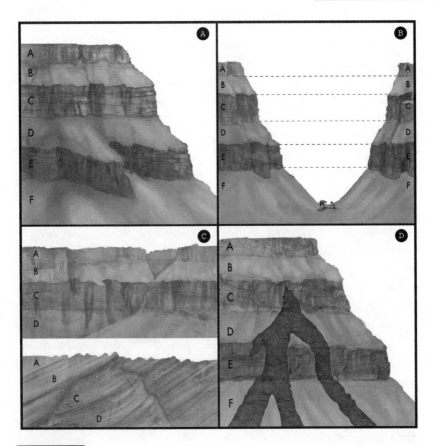

Figure 5.1 ▲

Steno's laws are used to determine the relative age of one rock body compared to another. (A) The principle of superposition says that the rocks near the top of a stack of layered sediments or lava flows are younger than those at the bottom of the stack. Thus Bed A is the youngest, and Bed F is the oldest. (B) The principle of original continuity says that rocks that match from one outcrop to another were once connected and have since been carved away by erosion. (C) The principle of original horizontality points out that rocks form in horizontal layers, so if you find them tilted or folded or faulted, the deformation is younger than the rocks it deforms. (D) The principle of cross-cutting relationship says that when a rock body (such as a dike of molten lava) or a fault cuts through another rock, the material that cuts through is younger than whatever it cuts. (Illustration by Mary Persis Williams)

When Steno looked at matching rock units exposed by erosion, he also realized that they were once continuous and only later were cut through by erosion (fig. 5.1B). This is the *principle of original continuity*. In the Apennine Mountains of Italy where Steno got his insights, many of the rocks are tilted, faulted, and folded. Steno realized that they had not been created that way but had been laid down horizontally, then later tilted and folded, so the deformation must be younger than the rocks it deforms (fig. 5.1C). This is the *principle of original horizontality*.

Another useful concept is the *principle of cross-cutting relationships*. If you have a molten igneous rock intruding into another rock (such an intrusion is usually called a "dike"), the rock that does the intruding must be younger that the rocks that it cuts through (fig. 5.1D). You can't cut through something if it isn't already there. Likewise, if a fault cuts through rocks, it must be younger than the rocks it cuts. These principles of relative dating first described by Steno in 1669 were widely in use by the time modern geology was born, about 1800 to 1830, and the geologic time scale was drawn. The various names for the eras and periods and epochs of the geologic time scale are relative ages (fig. 5.2).

The other way to date rocks is known as *numerical dating* (incorrectly called "absolute dating" in older books). This technique provides a date in number of years such as thousands of years or millions of years. Numerical dating is a young technique, developed in the early twentieth century, and the most popular method—potassium-argon dating—has only been around since the 1950s.

Numerical dating is done by measuring the ticks of the radioactive "clock" in certain minerals. As minerals crystallize out of a magma, they trap radioactive elements like uranium-238, uranium-235, rubidium-87, or potassium-40. These radioactive elements are naturally unstable, and they spontaneously decay into different elements. As this decay proceeds over millions of years, the unstable radioactive parent atoms decay into a stable form known as daughter atoms, such as lead-206, lead-207, strontium-87, and argon-40 (respectively, for each of the elements listed above). The rate of this decay is precisely known, so we can obtain the numerical date by measuring the ratio of parent atoms to daughter atoms in a crystal of a mineral such as feldspar or mica or zircon since that crystal was first formed.

This process only occurs in crystals that form from a molten rock, so we can only date igneous rocks directly. What about sedimentary rocks, which contain the fossils? We cannot directly date them by radioactive minerals. Instead, we need to find places where igneous rocks (such as lava flows or volcanic ash deposits) are interbedded with fossiliferous sedimentary rocks. If a bed has Oligocene fossils ("Oligocene" is a relative age term), and

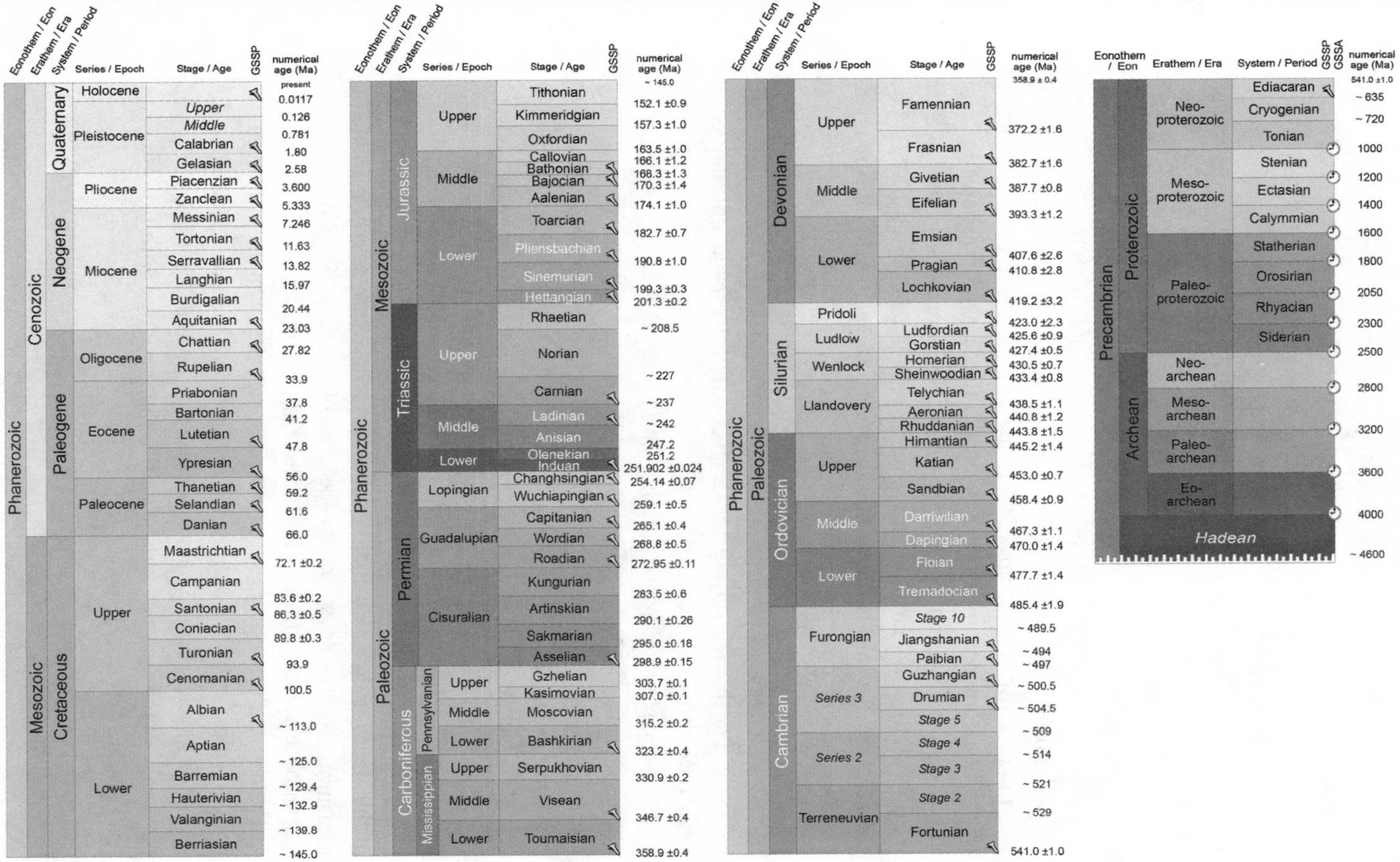

Figure 5.2 ▲

The modern geologic time scale. (Courtesy of the International Commission on Stratigraphy)

the bottom of the bed has an ash dated 34 million years old, and the top of the bed has a lava flow dated 23 million years old, then we bracket the age of the Oligocene between 23 and 34 million years ago. The entire geologic time scale was constructed this way: by finding fossiliferous sequences with fossils that gave well-determined relative ages, and then using any and all available igneous rocks that are in the right position to tell us the age.

There is one other radiometric system, known as *radiocarbon dating*, or carbon-14 dating. Unlike the other methods, you can date the fossil bones or shells or wood or any carbon-bearing substance directly because you are measuring the decay of unstable radioactive carbon-14 incorporated into the organism before it died. The main drawback of this technique is that radiocarbon decays very rapidly. Half of the original carbon-14 parent atoms are gone in just over 5,000 years, so the entire clock "runs down" in 60,000 to 80,000 years, and anything older than this cannot be dated by radiocarbon. This method is primarily used by archeologists who are interested in dating human bones and artifacts, and by paleontologists studying the last Ice Age, which spanned the interval from 80,000 to 10,000 years ago. It is useless to anyone studying older fossils because the clock is dead for them. For this reason, we could never date a Mesozoic dinosaur bone using radiocarbon.

The principles of numerical dating have told us that Earth is immensely old. Numerous meteorites and moon rocks are 4.6 billion years old, and that is how we estimate the age of the solar system. So far, the oldest Earth rocks are only 4.28 billion years old, and the oldest Earth minerals are 4.4 billion years old, so we have no Earth rocks as old as the moon rocks or meteorites. But this is not surprising because the earth's crust is constantly being mobilized and remodeled by plate tectonics. We do not expect that any crust from the oldest earth could survive. The oldest known fossils are bacteria from South Africa and Australia, which are about 3.4 to 3.5 billion years old, and organic carbon has been found in rocks 3.8 billion years old, which most scientists think is evidence of ancient life.

From these principles, we can reconstruct the sequence of events in any part of the earth. To do this, we need to know the exact position in the sequence of rocks where each fossil was found. And, if we are using lava flows or volcanic ash beds, they must be interbedded with fossils whose positions are precisely known.

COLLECTING FOSSILS

The best areas for collecting fossils are where the landscape is dry and barren, with no plant cover to obscure the rocks underneath. Typically, these areas form deeply eroded, weird landscapes known as *badlands* (fig. 6.1A–B). Badlands get their names because they are a bad place to lose a cow, or to try to cross if you don't have lots of water—but to a fossil collector, badlands are good lands. In fact, most of the best fossil collecting is found in areas that were given diabolical or hellish nicknames, such as "Devil's Punchbowl" and "Hell's Half Acre." They may have looked barren and forbidding to early settlers who were seeking flat grassy soils, but they are heavenly for fossil collectors.

The badlands are good lands to look for fossils that are extremely rare. Fossils of dinosaurs and prehistoric mammals and most backboned animals rarely occur in large concentrations such as shell beds or leaf beds. You need to do a lot of prospecting: walking along the exposed area slowly, eyes to the ground, looking for any telltale color or texture that tells you that you are looking at fossil bones or fossil teeth. Fossil bone, for example, can usually be recognized by its spongy texture on the inside and a solid outer core of bone, and often by a shape that corresponds to a known bone shape. For this reason, a good paleontologist is an expert in recognizing almost any bone of any kind of animal found in the fossil beds where the search takes place. Fossil teeth, which are the most durable and diagnostic parts of any mammal skeleton, are usually spotted by the shiny glint of enamel on their outer surface.

Most often, these fossils are collected after hundreds of hours walking slowly in the hot sun, trying to see the bones or teeth on the ground amidst

Figure 6.1 ▲

Two fossiliferous badlands exposures: (*A*) Oligocene rocks in Badlands National Park, South Dakota; (*B*) Miocene rocks in Redrock Canyon State Park, California, a filming location that claimed to be Cretaceous beds from "Snakewater, Montana" in the first *Jurassic Park* movie. (Photographs by the author)

all the other colored pebbles and other rocks that distract you from what you are seeking. Often you spot something that looks promising, only to pick it up and find it's just another pebble, or a concretion that looks like a fossil. In places where dinosaur or mammal fossils are found, the terrain can be very remote and rocky and even dangerous. Typically, badlands are steep and rocky. You must be prepared for hardcore hiking: good boots, comfortable durable clothes suitable for the weather, a broad-brimmed hat to keep the sun off and lots of sunscreen, and plenty of water and snacks in your pack, especially if you are out collecting all day in the heat and glare of the sun.

Once you spot a promising object sticking out of the ground, you might investigate it further to see if it's a fossil or not. Even if it does turn out to be bone, it might be too fragmentary or incomplete to be identified. If it's a worthwhile find, you may need a brush to clean it off, a rock hammer or an awl to pry it from the rock, or a chisel or pry bar to split open the rock to see inside (fig. 6.2). In addition, to collect certain kinds of fossils, it's wise

Figure 6.2 ▲

Collecting the giant fossil fish *Xiphactinus* from the Cretaceous chalk beds of western Kansas. (Photograph by the author)

to have a bag to put the fossil in (preferably a plastic bag with a ziplock top) and a felt-tipped pen or marker to label the bag. If the specimens tend to be fragile, you might also carry a roll of toilet paper to wrap the specimens in before bagging them up. To a paleontologist, toilet paper has many uses!

In addition, if the fossil is bone that has a tendency to shatter or break when exposed, most paleontologists carry a small plastic squirt bottle with a preservative or hardener. The early fossil collectors used substances like shellac, but most modern paleontologists use organic hardeners such as Glyptal and Alvar, which are much easier to clean off the specimen in the lab. Once the fossil is exposed, they squirt the hardener on the specimen and let it soak in and dry, gluing the fossil together in the field and making it less likely to break.

If the specimen is large and fragile, however, a much bigger effort is required. Serious paleontologists working on larger fossils usually carry some sort of material to make a plaster and burlap jacket around the fossil, similar to the plaster cast that a doctor puts on a broken arm. When a large fragile bone is found, the first job is to carefully uncover and partially clean the specimen until you find out how far back into the rock it goes. Once you have exposed its top surface completely, you dig a trench around it and dig away all the encasing rock until the fossil is sitting with a narrow pedestal of rock beneath it (fig. 6.3A). All through this operation, the exposed parts of the fossil are squirted with hardeners to prevent it from shattering.

When the large fossil is completely isolated on a pedestal of rock, typically you cover all the freshly exposed surfaces with a layer of wet toilet paper to protect it from the plaster. Then you mix up the plaster of paris in a small tub or bowl according to the package directions. Meanwhile, you cut burlap sacks into narrow strips. Each strip is then dipped into the plaster until it is saturated, then it is wrapped around all of the exposed surfaces of the specimen that are covered with wet toilet paper (fig. 6.3B). The specimen is plastered across the top and around the sides and along the exposed base. After a few hours, the plaster jacket will have dried enough that you can do the trickiest part of the process. Using pickaxes and shovels, the crew undermines the pedestal as far as they dare, then uses a pick or a long piece of wood as a lever to break the fossil off its pedestal and turn it over (fig. 6.3C). If the fossil doesn't shatter in the process (a tragedy that happens too often, even with the most careful preparations), then the exposed base of

Figure 6.3 ▲

Large fragile fossils may need a plaster jacket to protect them before they are transported. (*A*) This brontothere jaw and skull were originally spotted when only a tiny corner of the jaw was seen sticking out of the ground. The paleontologists then exposed the fossils until they were on a thin pedestal of sediment. (*B*) After exposing the fossils, a cast made of burlap soaked in plaster of paris was applied to the exposed bone to create a hard jacket to protect the fossil. (*C*) Once the plaster has dried, the paleontologist uses a pick to cut down the pedestal further, then pries it loose from the ground and flips it over, exposing the rest of the fossil. A plaster cast is then placed over the rest of the specimen to protect it during transport. (Photographs by the author)

Figure 6.3 ▲
(*continued*)

the fossil can be trimmed of excess rock. At this point, the plaster jacket can be completed over the exposed area, so the specimen is entirely encased in a hard plaster cast. Once all the plaster has dried, it can be carried out or loaded into whatever vehicle (a truck, or sometimes even a helicopter) is available and transported to a laboratory where it can be carefully cut out of the jacket and prepared.

COLLECTING FOSSILS

Most people don't happen to live near a desert region with good badlands exposures nearby. If you live in an area with a wetter climate and lots of vegetation, your fossil hunting options are more limited. When plants cover nearly every exposure, look for places where something cuts through the plants and soils and reveals the rock underneath. In much of the world, the roadcuts formed when highways are blasted through bedrock can be excellent places to collect. In areas near the sea, coastal cliffs and the beach below them are constantly eroded and scoured by the pounding of waves and the rush of the tides. If you walk on the beach during a very low tide (you can look up the tide tables online), you can find the treasures left by the ocean (fig. 7.1A).

In some cases, the fossils are literally everywhere, and you only need to reach out and pick them up (fig. 7.1B). There are many examples of fossil shell beds that are almost solid fossil shells, so finding them is no problem—you just have have decide what you want to collect and how much you want to carry (fig. 7.2A–C).

Figure 7.1 ▲

(A) Calvert Cliffs, with (B) shell
beds rich in Miocene molluscs.
(Courtesy of S. Godfrey)

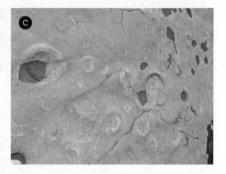

Collecting in the Lower Jurassic beds just west of Lyme Regis, on the Jurassic Coast of Dorset, England. (*A*) Exposures of the Lower Jurassic Lias beds of shale and limestone. (*B*) A large ammonite exposed in a boulder. (*C*) The bedrock surface exposed at low tide is full of eroded ammonites. (Photographs by the author)

COLLECTING FOSSILS

In many parts of the world, vegetation covers nearly all the rock, and there are few or no natural outcrops where you might find fossils. In these regions, paleontologists and geologists are forced to use whatever artificial exposures and outcrops are created by road building, construction, and quarrying. These exposures don't give us as much to work with as the miles of bare rock in badlands, or the miles of coastal cliffs, but they have to do. Some are surprisingly fossiliferous, so you don't need to cover a huge area to find specimens (fig. 8.1).

Quarries and roadcuts are limited exposures, and there are challenges associated with both of them. In the case of roadcuts, the biggest problem is traffic, especially if the road is very narrow. You have to be extra careful working in many roadcut exposures, and it's often a good idea to wear brightly colored vests or hats in safety orange so you are more visible to careless or inattentive drivers. Most of the time, collecting at the base of the roadcut is fine because fossils often fall down during the rains and accumulate at the base. If you must collect higher up, you should be extra careful about climbing up the roadcut, especially if you might fall down into oncoming traffic. It's always a good idea to watch out for rocks falling from above as well. Many collectors who work steep roadcuts wear hard hats for protection from falling rocks. Most roadcuts are public access and controlled by the local highway department or state agency that maintains the roads, so permission is not usually an issue. Some, however, are on private land, and you must obey all signs, especially "No Trespassing" signs.

Figure 8.1 ▲

A roadcut in northern Kentucky, just south of Cincinnati, Ohio. The layer the student is looking at is filled with large fossil corals that are eroding out. (Photograph by the author)

Some of the same problems also apply to quarries (fig. 8.2). Many are steep and unstable, so be careful when working at the base of the cliffs and wear a hard hat. If it's an active quarry under private ownership, you must get permission from the owner, and safety laws may require that you wear not only a hard hat but also steel-toed boots (as the workers must wear). In addition, most quarries under private ownership will require you to sign a liability waiver, so you cannot sue them if something happens to you on their property. Heavy equipment may be driving around in active quarries, so you have to be extra careful. A giant dump truck or earth mover as big as a building cannot stop on a dime if you step in front of it. Some quarries also have active blasting times, so you must heed any warnings from the workers and be wary if you hear the warning siren before the blast. Even abandoned quarries with no active legal owners can be hazardous because they are no longer maintained. They may be prone to collapse and cave-ins after long periods of neglect, and you should be extra careful in these settings.

Finally, both quarries and roadcuts have the same problem in humid regions: plants can grow back very quickly if the quarry or roadcut isn't constantly cut back. In many places, the highway department makes beautiful roadcut exposures, only to immediately cover them up with various

Figure 8.2 ▲

Collecting in a rock quarry. (Photograph by the author)

forms of plants and stabilization materials so the roadcuts don't collapse onto the roadbed. In the southeastern United States, collectors have come to hate kudzu, a plant that grows over almost anything if not actively cut back; in the southwest, road crews stabilize roadcuts with drought-resistant ice plants. In parts of the country where rapid active construction for new housing tracts and buildings exposes clean rocks and fossils, you will have only a few months from the time the exposures are made to hunt for fossils before they are covered up again.

If you are an active collector, it is a good idea to join any local fossil clubs in your region. Club members often have the latest information on whether an exposure is good or grown over, where fresh roadcuts have just been made that will soon vanish, and who owns the property rights to important collecting sites.

THE CRUCIAL STEP

I have described where to look for fossils, and what to do when you find them, but the most important step of all is to always record exactly where you found the fossil and what rock unit and level it came from. If you are just collecting lots of shells from well-known localities where hundreds of specimens are already known (and many more remain to be found), this step isn't crucial. Still, it's good practice never to leave the field with specimens without noting the locality information (fig. 9.1A–B). If you don't do it while you are there, you are unlikely to accurately reconstruct your exact location after the fact.

Location information can be found in a number of ways. Lots of mobile devices have GPS systems built into them, so you can find your exact latitude and longitude if you know how to use the right app. In remote badlands, however, you may not have a cell signal, so it's wise to carry a portable GPS device that can pick up satellites anywhere. However, if you find yourself in a dense forest or in a narrow canyon, even satellite reception may be difficult. It is a good idea to have an "old school" backup system: a topographic map of the region. This map will help you navigate around, but it is especially useful in locating and recording the exact spot where the fossil was found.

In addition, for the specimen to have any scientific value, you need to provide as much detail about the geologic location of the fossil as possible, including the exact formation name the level within the formation where it was found. For a rigorous scientific study, all the fossils must have their exact level recorded from a particular top or bottom of the formation or a marker bed within the formation (figure 9.2A–B).

Figure 9.1 ▲

(A) A paleontologist taking notes on an outcrop while the crew collects fossils. (B) A close-up of a field notebook showing the details of the fossil locality, sometimes with a Polaroid photo attached. (Photographs by the author)

This may seem like a big hassle, but it is a crucial step that is needed for a fossil to have real value to science. Far too often I've run across fossils in museums that had only very vague locality information. They may make pretty mantelpiece displays, but they are of no use if you want to do any kind of study that depends on precise locality information. Many paleontologists have been shown a very important and age-diagnostic fossil to identify, but when asked the collector can't remember where it came from—the value of the fossil is greatly diminished as a result. The fossil could have unlocked clues or solved important geologic problems, but without the location information, the paleontologist can do little except identify it.

Dr. Robert J. Emry, a mammalian paleontologist now retired from the Department of Paleobiology at the Smithsonian Institution, told me a revealing story when I was his field assistant in 1978. In the 1950s and 1960s, he was working in the field crew of the legendary Dr. Morris Skinner of the Frick Laboratory of the American Museum of Natural History in New York. When he was a teenager and first began collecting,

Figure 9.2 ▲

Every specimen must have its precise geographic information and stratigraphic level recorded when it was collected in the field, or its scientific usefulness is limited. (*A*) A museum display of Cretaceous sea urchins from the Dover Chalk, each with precise locality information on the labels. (*B*) Two paleontologists doing research on fossil camels in the American Museum of Natural History in New York. Each specimen in the tray has a catalog number and all the geographic and stratigraphic information displayed on the specimen or on the label in the tray. (Photographs by the author)

Bob found a fossil skull and rushed over to Morris with his find. Morris asked him if he knew precisely at what level he had found it. When Bob didn't recall, Morris just tossed it away because its scientific value was so worthless that it wasn't even good enough to take home.

The most important reason for recording the exact level in the formation is to help date the formations from which fossils come. In the 1790s, British canal engineer William Smith noticed that every formation has its own distinct assemblage of fossils. This is known as the *principle of faunal succession* (also called "fossil succession"). He discovered that he could tell each formation apart from the others by its fossil content. Soon he impressed the rich gentlemen collectors by telling them what formation they had gotten their fossils from. Building on this discovery, Smith embarked on an ambitious project to map all the rock formations of Great Britain (fig. 9.3). Simon Winchester called it "the map that changed the world" because a geologic map is the fundamental tool for understanding Earth's history. It is also a tool for finding oil, gas, coal, and other resources.

Smith was just a humble engineer, and not a rich gentleman, and he ended up going bankrupt doing this enormous project, even as other geologists stole credit for his discoveries. He was even put in debtors' prison for a while, but nevertheless he published his masterpiece in 1815, the first real geologic map ever produced. Fortunately, late in his life other geologists began to acknowledge his contributions to the discovery of Earth's history and geology. In 1831, shortly before he died, Smith was hailed as the "Father of British Geology."

So how do we use faunal succession to tell time? The method of dating rocks using fossils is called *biostratigraphy*, the study of the layered sequence of rocks and their fossils. First, the geologist or paleontologist makes careful note of the exact level in the sequence of strata from which every fossil comes. When these specimens are plotted on a stratigraphic column (fig. 9.4), they show a distinct sequence of different fossils marking different ranges in time (and in the strata). This is called a *range zone*. The paleontologist then does the same careful recording and plotting of biostratigraphic data and range zones on other stratigraphic columns. When the columns are compared, the sequences can be matched, or correlated, between the two different areas.

For example, in figure 9.4, the sequence of oysters, belemnites, and snails allows a straightforward match between area A and area B. In area C, however, the belemnites and the oysters are found at the top of the column, so that part of column C is correlated with the bottom of columns A and B. This establishes that the range zone with the sea star fossils must be older than the zone with the oysters, and it gives us a longer composite sequence. Finally, area D has a different fossil sequence. The belemnites and snails allow us to match the bottom of column D with the top of columns A and B. The presence of the fish fossils in column D suggests that there may be a range zone missing between the snails and the belemnites in columns A and B. Finally, the ammonite zone appears above the snail zone, giving an even longer composite sequence. Thus the overall sequence of fossils and time goes from the oldest (sea star fossils) zone to the oyster zone, then to the belemnite zone, the fish fossil zone, the snail zone, and finally the ammonite zone.

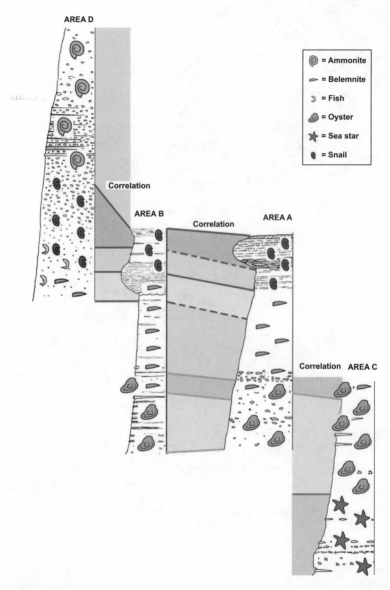

AREA D

AREA B

AREA A

AREA C

Correlation

Correlation

Correlation

= Ammonite

= Belemnite

= Fish

= Oyster

= Sea star

= Snail

Figure 9.4 ▲

Stratigraphic columns showing correlation between four local range zones to form a composite biostratigraphy of the entire time interval. (Illustration by Mary Persis Williams)

When I did my dissertation project on the mammal fossils of the White River Badlands of the Dakotas, Nebraska, Colorado, and Wyoming, most of the fossils I found in older museum collections were useless for the detailed type of study I had planned. Luckily, Morris Skinner and his collectors had made enormous collections of Badlands fossils in all these areas. Unlike any previous group of collectors, they had recorded the exact stratigraphic position of every fossil to the nearest foot above or below a marker bed. This allowed me to precisely date the fossils and to do a detailed study of their change through time—which would not have been possible if the information had not been recorded in the field.

Knowing exactly where you found the fossil and recording it before you leave raises another issue: the land and collecting rights. If you are collecting in roadcuts or at beach cliffs, these public access lands usually have no restrictions. But if you go into a quarry or onto any kind of private land, you have to obtain permission to collect from the landowners, and they may have restrictions on what you can take home as well. If you are collecting on any kind of federal lands (national parks, national forests, national grasslands, or anything managed by the Bureau of Land Management), or any kind of state lands (state parks, state forests, state monuments, etc.), you need a formal permit before collecting. The procedures for getting those permits vary from place to place, but you should not start walking across publicly managed lands and grabbing fossils without first getting the proper permits.

This is a common problem. Careless amateurs and professional poachers often take fossils from publicly held lands—of which you as a taxpayer are the part owner—and sell them for a profit or keep them as a treasure without obtaining the legal rights to do so. Sadly, too often important scientific specimens have vanished into rich people's living rooms because poachers and careless amateurs break the law—and the agencies who manage the land don't have the people or money or other resources to patrol these lands and catch the poachers. Once such specimens vanish into private hands (often without recording any of the proper locality information or level within the formation), they are lost to science, and their most important data are lost forever as well.

PART II

IDENTIFYING YOUR FOSSILS

WHAT'S IN A NAME?

Once a fossil is found, and its locality information recorded, the next step is determining what it was. We assign scientific names to fossils just like we do for living animals and plants. Scientific names may seem a bit long and hard to pronounce, but they are essential to scientific communication. The popular or common name of many living animals and plants differs from culture to culture and language to language. For example, a peccary to English speakers is a *javelina* in Latin America, and a lion to us is *simba* to Swahili speakers. Even within the same language, the common name may not be consistent. If you say "gopher" in some parts of the United States, it means a small burrowing rodent, but in other parts it means a gopher tortoise.

For this reason, every organism (plant, animal, fungus, and even bacteria) has its own scientific name (fig. 10.1). Scientific names are universal around the world, no matter what language the scientist speaks. You may not be able to read much of a scientific paper written in Mandarin Chinese, but the scientific names are always printed out in Roman script, so anyone can read them and at least guess what animal is the subject of the research. The scientific name for the burrowing rodent some people call a "gopher" is *Thomomys*, but the gopher tortoise is *Gopherus*, so there is no confusion.

It is essential to know the scientific name for fossils because most fossils don't even have a common name. You may know the saber-toothed cat by its English name, but it's different in other languages—and to all scientists it is *Smilodon*. Mammoths are familiar to us by that name, but in other languages they could be *mamut* (Spanish) or *mammouth* (French), but they are all *Mammuthus* to a scientist. Most fossil animals and plants have no

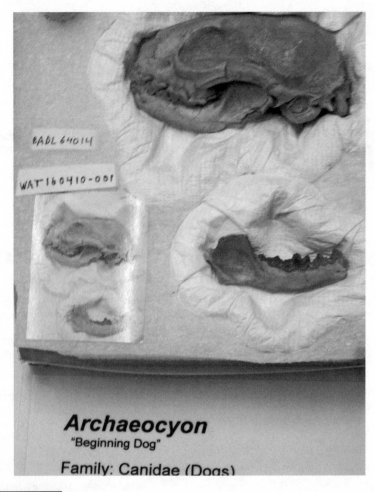

rADL 64014

WAT 160410-001

Archaeocyon
"Beginning Dog"

Family: Canidae (Dogs)

Figure 10.1 ▲

Fossil of an early dog from the Big Badlands of South Dakota showing the catalog number and the scientific name *Archaeocyon* in italics, and a translation of what its name means. (Photograph by the author)

common name whatsoever, so there is no choice but to use their scientific names. You actually know quite a few scientific names of prehistoric and living creatures. For example, everyone knows *Tyrannosaurus rex*; that is its proper scientific name, and no other popular name exists. Nearly all of the other dinosaur names you probably know—from *Brontosaurus* to *Velociraptor* to *Stegosaurus* to *Triceratops*—are scientific names as well.

All organisms on Earth have a two-part (binomial) scientific name. The first part is the genus name (the plural is "genera" not "genuses"). It is always capitalized and either italicized (in print) or underlined (when hand-written). The names *Tyrannosaurus, Brontosaurus, Velociraptor, Stegosaurus,* and *Triceratops* are all genus-level names or "generic names." But a genus typically includes a number of species. The species name (or "trivial name") is never capitalized (even if it came from a proper noun), but it is always underlined or italicized. Thus *Tyrannosaurus rex* is a genus and species name; so is *Velociraptor mongoliensis.* Your scientific binomen is *Homo sapiens,* but there are other species of *Homo,* such as *Homo neanderthalis, Homo erectus,* and *Homo habilis.*

Generic names are never used more than once in the animal kingdom. There are a few cases of the same genus being used for both plants and animals, but there is no likelihood of confusion between a plant and an animal. Species names, however, are used over and over again, so they cannot stand alone in a scientific paper. Thus you can say *Tyrannosaurus rex* or *Homo sapiens,* but not "rex" or "sapiens." You can abbreviate the genus name, so *T. rex* or *H. sapiens* is proper.

Scientific names were originally based on Latin or Greek words because all scholars read and wrote in Latin or Greek as an international form of com-munication in the early days of natural history. Thus most scientific names can be broken down to their original meaning. *Tyrannosaurus rex* means "kind of the tyrant lizards" and *Homo sapiens* means "thinking human."

The criterion of Greek or Latin roots and latinization of names has become more relaxed as fewer and fewer scientists learn the classical languages (the standard languages for all scholars less than a century ago), and much work is now being done in China, Japan, Russia, India, Latin America, and other non-European scientific communities. Scientists have gotten more and more creative with naming, often to the point of silliness or to selecting names that are difficult for others to use. For example, in 1963 mammalian paleontologist J. Reid Macdonald gave names based on the Lakota language to a number of specimens recovered from the Lakota Sioux reservation land near the old site of the Wounded Knee Massacre in South Dakota. Most non-Lakotans find them difficult to pronounce or spell. Try wrapping your tongue around *Ekgmowechashala* (iggi-moo-we-CHA-she-la), which means "little cat man" in Lakota. It is a very important specimen of one of the last fossil primates in North America, so it has gotten a lot of attention, and many people have struggled to pronounce its name. In the same paper, Macdonald also named *Kukusepasatanka*, a hippo-like anthracothere; *Sunkahetanka*, a primitive dog; and *Ekgmoiteptecela*, a saber-toothed carnivore. Then there is the transitional fossil between seals and their ancestors known as *Puijila*, which comes from the Inuktitut language of Greenland; you need to go to the website (http://nature.ca/puijila/fb_an_e.cfm) to hear the correct pronunciation. In Australia, many fossils have odd-sounding names with Aboriginal roots, such as *Djalgaringa, Yingabalanaridae, Pilkipildridae, Yalkparidontidaem, Djarthia, Ekaltadeta, Yurlunggur, Namilamadeta, Ngapakaldia,* and *Djaludjiangi yadjana.* Some others include *Culmacanthus* ("culma" is Aboriginal for "spiny fish"), *Barameda* (Aboriginal for "fish trap"), and *Onychodus jandamarrai* after the Jandamarra Aboriginal freedom fighters. *Barwickia downunda* is named after Australian paleontologist Dick Barwick. *Wakiewakie* is an Australian fossil marsupial, supposedly named from the Australian way of waking up sleepy field crews in the morning.

About a century ago, an entomologist named Kirkaldy got a bit too creative naming difference genera of "true bugs," or Hemiptera. He published the names *Peggichisme* (pronounced "peggy-KISS-me"), *Polychisme* for a group of stainer bugs, *Ochisme* and *Dolichisme* for two bedbugs, *Florichisme* for a plant hopper bug, and *Marichisme, Nanichisme,* and *Elachisme* for seed bugs. For leaf hoppers and assassin bugs, Kirkaldy used male names such as *Alchisme, Zanchisme,* and *Isachisme.* In 1912 the Zoological Society of London officially condemned his naming practices, although as long as they were valid taxa, they could not abolish the names.

An entire website devoted to weird names (http://www.curious taxonomy.net/) lists the gamut of odd inspirations, from puns to wordplay to palindromes that read the same way forward and backward. Some of the more clever names include the clams *Abra cadabra* and *Hunkydora*; the beetle *Agra vation*; the snails *Ba humbugi* and *Ittibittium* (related to the larger snail *Bittium*); the flies *Meomyia, Aha ha,* and *Pieza pi*; the wasps *Heerz tooya* and *Verae peculya*; the trilobite *Cindarella*; the Devonian fossil *Gluteus minimus*; the fossil carnivore *Daphoenus* (pronounced Da-FEE-nus) *demilo*; the fossil snake *Montypythonoides*; the Julius Caesar-influenced extinct lorikeet *Vini vidivici* and the water beetle *Ytu brutus*; and the Australian dinosaur *Ozraptor* (known as the "Lizard of Aus"). After a few too many beers, paleontologist Nicholas Longrich named a horned dinosaur *Mojoceratops* because it had an elaborate heart-shaped frill that might have improved its ability to attract mates. There is a Cretaceous lizard named *Cuttysarkus* (revealing the namer's preference for that brand of Scotch whiskey) and a dog-like fossil mammal known as *Arfia*. The oldest known primate fossil is *Purgatorius*, not because the namer had some sort of religious point to make about humans but because it was found in Purgatory Hill in the Hell Creek beds of Montana (suitably hellish in the summertime with hot temperatures and dangerous slopes). There are also fossils named after characters in the Star Wars, Lord of the Rings, and the Harry Potter series. Despite the musty reputation of taxonomists working away in dark museum basements, they certainly have a sense of humor!

Taxonomic names sometimes attempt to describe the creature or give some idea of its main features. However, if the name becomes inappropriate, it is still valid as long as no other senior synonyms are known. For example, the earliest known fossil whales were originally mistaken for large marine reptiles and named *Basilosaurus*, or "emperor lizard." Only later did scientists realize they were whales and mammals, not lizards, but the name is still valid even though it is inappropriate. In the 1920s scientists retrieved material of a bizarre predatory dinosaur from the Cretaceous of Mongolia and named it *Oviraptor* ("egg thief") because of its proximity to nests of eggs they thought belonged to the most common dinosaur there, the horned dinosaur *Protoceratops*. But in the 1980s and 1990s, expeditions returned to Mongolia and found fossil skeletons of *Oviraptor* mothers brooding those same eggs, and the bones of unborn *Oviraptors* inside the eggs. The "egg thief" was actually the *parent* of the eggs, not a thief at all, but this slander to *Oviraptor* cannot be changed just because it's now inappropriate.

In addition to names with difficult, odd, or funny pronunciations and meanings are names that honor individuals as well as a tick or a leech or some other parasite that is named after people they wished to *dishonor*. Even though the International Commission on Zoological Nomenclature (ICZN) has a clause stating that "no zoologist should propose a name that, to his knowledge, gives offense on any grounds," the rule has been violated many times. Linnaeus himself named a noxious weedy aster *Sigesbeckia* after his rival Johann Sigesbeck, who opposed Linnaeus's sexual classification of plants. A zoologist named a piranha *Rooseveltia natteri* because he hated President Theodore Roosevelt. Three different species of slime mold beetles are named after former President George W. Bush, Vice President

Dick Cheney, and Defense Secretary Donald Rumsfeld. There is a species of louse named after the famous *Far Side* cartoonist Gary Larson (*Strigiphilus garylarsoni*), although this was intended to honor, not dishonor, him (and reportedly Larson loved it). The famous late-nineteenth-century paleontologists Edward Drinker Cope and O. C. Marsh insulted each other with naming wars. Marsh named a marine lizard *Mosasaurus copeanus* (emphasis on the last four letters), and Cope named a fossil hoofed mammal *Anisonchus cophater* (emphasis on the last five letters). Cope told his protégé Henry Fairfield Osborn, "Osborn, it's no use looking up the Greek derivation of cophater, . . . for I have named it in honor of the number of Cope-haters who surround me." A century later in 1978, Leigh Van Valen returned the compliment by naming another primitive hoofed mammal after Cope: *Oxyacodon marshater*. The huge pig-like mammal *Dinohyus hollandi* was named by paleontologist O. A. Peterson after Carnegie Museum director W. J. Holland, who put his name as first author on almost every paper even if he didn't do the research or write any of it. The name means "Holland's terrible pig." When the specimen was announced in the Pittsburgh newspaper, the front-page headline was "*Dinohyus hollandi*, The World's Biggest Hog!"

The species is the fundamental unit in nature because it is species that evolve due to natural selection on populations within the species. The genus is a bit more arbitrary, depending on the scientists' judgments as to which species cluster together. Genera are clustered into larger groups known as families. For example, our genus *Homo* belongs to the family Hominidae, along with other genera such as *Sahelanthropus, Ardipithecus, Paranthropus*, and *Australopithecus*. The dogs are all members of the family Canidae, the cats are Felidae, and the rhinoceroses are in the Rhinocerotidae. In the animal kingdom, all family names end with the suffix "idae," which is a quick clue when you encounter an unfamiliar name. (In the plant kingdom, family names end in "aceae," so Rosaceae is the plant family that includes roses.)

Families are clustered into a larger group called an order. Humans, apes, monkeys, lemurs, and their relatives form the order Primates, and the order Carnivora includes most of the flesh-eating mammals such as cats, dogs, bears, hyenas, weasels, raccoons, seals, and walruses. The rodents are an order (Rodentia), as are the rabbits (Lagomorpha), and most of the larger groups of mammals are orders. Orders are clustered into classes. Within the backboned animals, the families of mammals are clumped into class Mammalia, and the birds (class Aves), the Reptilia, the Amphibia, and so on are classes. Classes are clustered into a larger group called a phylum (plural is "phyla"). Vertebrates (animals with backbones) are members of the phylum Chordata, but there is a phylum Mollusca (molluscs, including clams, snails, squids, and their relatives), the phylum Arthropoda (jointed segmented animals, including insects, spiders, scorpions, crustaceans, millipedes, trilobites, and many others), and so on. The highest rank of all is the kingdom (fig. 10.2). We are members of the kingdom Animalia, but there are also kingdoms for the plants, the fungi, and so on.

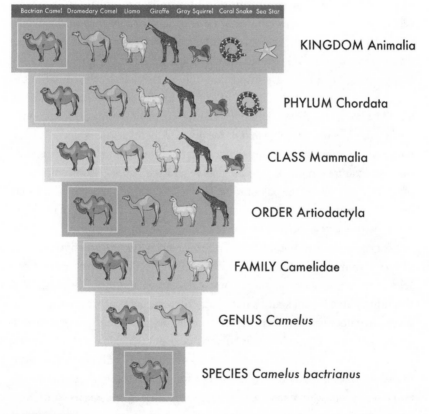

Figure 10.2 ▲

The hierarchy of classification showing how each rank or group is nested within a larger one. (Illustration by Mary Persis Williams)

Here is an example of how the hierarchy of groups within groups looks.

Kingdom	Animalia	Animalia (animals)
Phylum	Chordata	Mollusca
Class	Mammalia	Gastropoda
Order	Primates	Neogastropoda
Family	Hominidae	Turritellidae
Genus	*Homo*	*Turritella*
Species	*sapiens*	*ocoyana*

Strict rules for how organisms can be named are explained in official rule books, such as the *International Code of Zoological Nomenclature*, which used to be available only in bound printed copies but now can be read online (http://iczn.org/code). There are similar codes for plants, fungi, and bacteria and protists. Most rules are important only to specialists who are about to create a new scientific name, but other rules are commonly encountered by anyone who follows fossils or biology and are worth mentioning. Most important is the *rule of priority*. The first name given to an organism is the only valid name (unless there are problems), no matter how unfamiliar or inappropriate it is. For example, most paleontologists regard the name "Brontosaurus" as invalid because the same paleontologist, O. C. Marsh, who named that fossil, gave the name *Apatosaurus* to another specimen of the same animal earlier. Thus *Apatosaurus* is the proper senior synonym of "Brontosaurus," and paleontologists have been bound by this rule ever since Elmer Riggs figured it out in 1903. No matter how familiar the public is with the name Brontosaurus, scientists cannot use that name. (Some paleontologists have recently tried to revive the name Brontosaurus, but this is still controversial.)

When paleontologists are working on fossils, they have to keep track of all the names that have been given and figure out which name has priority; the other later names are known as junior synonyms. This is true even if the senior name turns out to be inappropriate. In the earlier part of the book, we saw how an early fossil whale was called *Basilosaurus* ("emperor lizard" in Greek) even though later work showed it was a whale and a mammal, not a reptile. By the rule of priority, *Basilosaurus* must stand no matter what it means.

In addition to rules about which name is valid, there are strict rules about creating new names for new species or genera. For the last century, a new scientific name must include a clear diagnosis of how to distinguish it from other similar species, a good description of the specimens, good illustrations, a list of specimens considered to be part of the species, a type specimen that is the conceptual basis for the species, the geographic range and time range of the species, and many other things. All of these must be published in a reputable scientific journal, not on a web page or unpublished dissertation or somewhere else. Otherwise, the new name of a genus or species is not valid, and other scientists will not recognize or use it. A scientist cannot name a genus or species after himself or herself, but the scientist can name it after someone else, and have that person return the favor on a different fossil. For example, my friend and colleague Spencer Lucas named a fossil rhinoceros *Zaisanamyndon protheroi* after me, and I recently named a fossil peccary *Lucashyus* in his honor. Some of my colleagues in South America honored me by naming a fossil peccary *Protherohyus catadontus*, and I've written a paper honoring them with a name on my next new species of peccary.

These rules may seem boring and excessively legalistic, but they are essential to maintain order and stability in scientific names. Scientists agreed to these rules over a century ago to prevent pointless arguments about whose name for an organism is right. All other scientists (and especially the scientific journals) follow these rules, and journals will not publish any work that violates them. It's like knowing the rules of the road before you take your driving test. The Department of Motor Vehicles, and all other drivers, must assume that you know the proper rules for driving because they don't want to be victims of a deadly accident if you suddenly break the rules. In many cases, amateur fossil collectors have tried to create new names, even to publish them in books and websites, without following the rules properly. The rule book allows professional scientists to quickly determine who is right and who is not, and whose work deserves attention and whose work ought to be ignored.

PHYLUM PORIFERA

Most people hear the word "sponge" and think of a block of synthetic foam, or of the cartoon character SpongeBob SquarePants. But sponges are actually the simplest and most primitive multicellular animals on the planet (fig. 11.1). Their fossil record goes back to the earliest Cambrian, and molecular data suggest that they originated even earlier but left no hard parts for many millions of years.

Sponges are just one step above single-celled organisms. Each sponge cell is completely independent of the others and performs all of its own biological functions such as feeding, breathing, excreting waste, and reproduction. They have no specialized tissues or organs, as all other animals do. They are loosely connected to each other on a structure (or skeleton) made of tiny, woven, needle-like pieces known as spicules, which the individual cells collaborate to secrete as a support. They are so independent that if you force a sponge through a fine sieve it will reassemble itself into a new sponge. I doubt you could do the same if you were forced through a sieve!

The basic sponge structure is a tube-like or conical structure, shaped roughly like a chimney. The simplest sponges have only a single thin wall (fig. 11.2A) made of interwoven spicules, punctured by many small holes or pores throughout its surface. This is where the phylum gets its name, Porifera or "pore bearing" in Latin. Individual sponge cells line the wall of the tube and especially the canals that connect the outer pores with the inner pores into the central cavity. These cells have a whip-like flagellum

Figure 11.1 ▲

A reef in the Cayman Islands that hosts a variety of living sponges, including the yellow tube sponge *Aplysina fistularis*, the purple vase sponge *Niphates digitalis*, the red encrusting sponge *Spiratrella coccinea*, and the gray rope sponge *Callyspongia*. (Courtesy of U.S. National Oceanic and Atmospheric Administration)

that drives currents past them, allowing them to trap tiny food particles and oxygen and release their waste products.

All the currents flow from the outside to the *spongocoel* cavity into the middle of the sponge and then out the top of the chimney (fig. 11.2B), known as the *osculum* ("little mouth" or "little kiss" in Latin). The one-way flow of water through the walls and out the top is propelled by the flagella. The flow is also enhanced by the fact that the top of the sponge has a weak suction that draws upward like a chimney. The water flowing over the top of the sponge must flow faster than the water around it, and this increased velocity means less pressure. The decreased pressure over the top of the sponge relative to the rest of the surrounding water forces the water through the pores and out the top. This is the "chimney effect." The top of a chimney

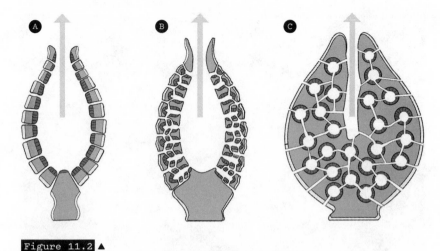

Figure 11.2 ▲

The basic structure of the three types of sponges. (*A*) The simplest is the ascon type, which is a simple, thin-walled cylinder that pulls in water from the sides through the pores and then out through the top (*see arrow*). The sponge cells (dark areas) line the inner surfaces of the sponge and capture their food and oxygen in those currents. (*B*) The sycon sponge is also built as a simple cylinder, but it has much thicker walls penetrated by canals lined with sponge cells (dark areas). (*C*) The most complex is the leucon sponge, which has much thicker porous walls and only a small spongocoel in the middle. (Illustration by Mary Persis Williams)

has negative pressure compared to its sides, which creates an airflow that goes up through the chimney.

Sponges are very efficient at passing water through their canals. The entire internal volume of a typical sponge is replaced with new water nearly every minute. A black loggerhead sponge about 50 centimeters in diameter and 30 centimeters tall may draw about 1,000 liters of water through its canals in a single day. Some sponges may flush the equivalent of 10,000 to 20,000 times their internal water volume in a single day. To see this action, Google the term "sponge currents" and find one of the many outstanding videos that show a diver releasing a harmless dye into the water just outside the wall of the sponge. Within a few seconds, a strong plume of dye-filled water pours out the top.

The simplest sponges are just thin-walled tubes or cones or long cylinders. But most sponges have much thicker walls made of many spongy, porous areas perforated by long sinuous canals (fig. 11.2C). These formed the bath sponges that were originally collected on the seafloor by divers before synthetic foam replaced them. Most fossil sponges have very simple structures, with a rough shape that can vary a lot if the sponge lived in different environments. Their spicules are also diagnostic, and individual tiny spicules are often found in deep sea sediment. Some make their spicules out of silica, and they are known as glass sponges (fig. 11.3). Others use the common mineral calcite, and they are known as calcareous sponges. The ones that were once used for bathing make their spicules out of the flexible organic material known as spongin, which allows their skeleton to compress and be squeezed without destroying it.

Sponge fossils may not be nearly as spectacular as trilobites or dinosaurs, but during certain places and times they were extremely important to the ecology of the seafloor. Many organisms, including various worms, arthropods, fish, molluscs, and protozoans seek shelter in sponges because of their large, hollow, protective spongocoel. Some sponge predators may eat sponges to get at the sheltered animals inside. A single black loggerhead sponge was reported to contain over 10,000 organisms within its canals and skeleton. In some regions, sponge fishermen find so many hard-shelled molluscs in the sponges that their catches are worthless as bath sponges. The delicate, glassy, Venus's flower basket sponge *Euplectella* is prized in the Orient as a wedding present (fig. 11.3). When shrimp move in and molt enough times, they become trapped because they are too large to escape through the grill over the osculum. Apparently, the pair of shrimp trapped in the glassy cage is symbolic of marriage in some cultures.

During the Paleozoic, sponges were the major reef builders in a number of instances, and they contributed to reefs built with corals as well. The first large colonial reef-building organisms on Earth were a group known as the Archaeocyatha (fig. 11.4). They formed huge reefs in the Early Cambrian in many parts of the world. They are organized in a structure similar to a sponge, with a cone-in-cone structure and an "I-beam" wall construction that separates the outer and inner walls. They were perforated by numerous pores, so they almost certainly fed and lived like sponges. However, paleontologists are still debating whether they were truly sponges or were only an early extinct experiment in the sponge-like body form. Archaeocyathans don't have spicules like all other sponges, and there are many other fundamental structural differences as well. The debate may never be resolved because archaeocyathans vanished completely by the Middle Cambrian when undoubted sponges replaced them.

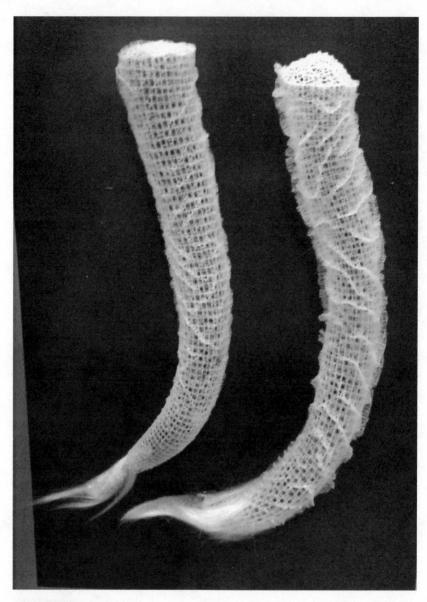

▲

Venus's flower basket sponge, *Euplectella*. With the soft tissue gone, the delicate interwoven basket of glassy spicules can be clearly seen. (Courtesy of Wikimedia Commons)

Figure 11.4 ▲

Sketch of a typical archaeocyathan, showing the porous conical outer wall with the double-walled internal structure, surrounding a central cavity. (Illustration by Mary Persis Williams)

Other sponges are easy to recognize and can often be good index fossils. A typical Silurian sponge is *Astraeospongia* (star sponge), so named because specimens have star-shaped spicules (fig. 11.5A–B). This sponge forms a flat, concave, dish-like body, with the spicules etched out during weathering. Another common Silurian sponge is *Astylospongia*, which has thick walls

Figure 11.5 ▲
Astraeospongia, the "star sponge": (*A*) the overall shape is like a thick-walled bowl or dish; (*B*) close-up of the surface, showing the interwoven star-shaped spicules. ([*A*] Illustration by Mary Persis Williams; [*B*] courtesy of Wikimedia Commons)

and narrow canals and is about the same size and shape as a golf ball (fig. 11.6A–B). During the Silurian and Devonian, huge reefs were built in the tropics consisting of corals plus a sponge group known as the stro-matoporoids (strohm-a-TOP-or-roids). These have a distinctive laminated

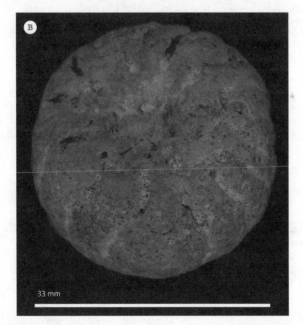

Figure 11.6 ▲

Astylospongia, a common Silurian sponge about the size of a golf ball: (*A*) drawing showing their overall shape; (*B*) photograph of a well-preserved example. (Courtesy of Wikimedia Commons)

structure in cross section (fig. 11.7A), with little bumps called mamelons, on the top surface (fig. 11.7B). Then there was a mass extinction in the Late Devonian, and stromatoporoid reefs vanished. They were replaced by reefs built by the glass sponge *Hydnoceras*, which were extremely common in the postextinction world of the latest Devonian (fig. 11.8A–B).

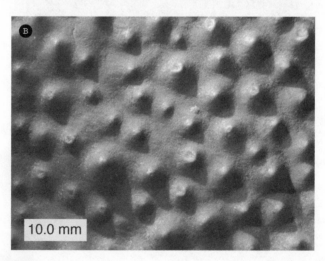

Figure 11.7 ▲

The layered sponges called stromatoporoids were the major reef-builders in the Silurian and Devonian, but they were wiped out during the Devonian mass extinction. (*A*) Their layered structure is very distinctive, as are (*B*) the tiny bumps (mamelons) on the top surface of each layer. (Courtesy of Wikimedia Commons)

Figure 11.8 ▲

The Devonian glass sponge *Hydnoceras*: (*A*) an individual sponge; (*B*) a colony of glass sponges from the latest Devonian. Cold-tolerant glass sponges replaced the coral-stromatoporoid reefs that had dominated the warm waters of the tropics in the Silurian and most of the Devonian. (Courtesy of Wikimedia Commons)

PHYLUM CNIDARIA (COELENTERATES)

Sponges are about as simple as a multicellular animal can be. They are just a colony of independent cells that secrete a shared skeleton; otherwise they are unspecialized and can perform all of their functions individually. The next more advanced level of organization is to have cells specialized for certain functions, such as an inner or outer layer of tissue arranged around a central body cavity. In addition, organ systems like the nervous system can be well developed, so there are discrete nerve cells. This is the level of organization you find in the phylum Cnidaria (called "Coelenterata" in older books). The roots of both of these names describe the group well. *Knidos* is the Greek word for "stinger" or "nettles," and all cnidarians have specialized stinger cells in their tentacles that are used to paralyze prey. The obsolete older term "Coelenterata" means "hollow gut" in Greek, which describes the fact that this phylum is the first and most primitive group of animals on Earth to have an internal body cavity (fig. 12A-D).

You are probably familiar with several kinds of living cnidarians, especially sea jellies (formerly called "jellyfish," but they are not true fish) and sea anemones. All cnidarians are built around the same body plan. They have a body arranged around an internal cavity, with an opening that serves as both mouth and anus at one end, and tentacles with stinger cells arranged around the mouth/anus. The internal wall of the body cavity is lined with *endodermal cells*, which are specialized for digesting prey that they swallow, and the external surface of the body is covered by *ectodermal cells*, which protect it from the outside world. They may also have masses of tissues between the endoderm and ectoderm called mesoderm or

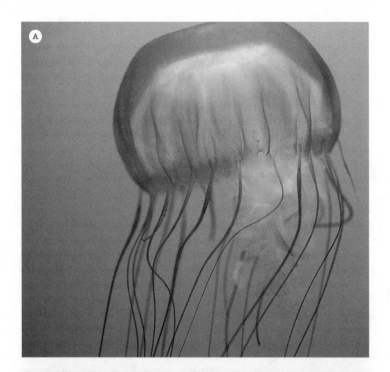

Figure 12.1 ▲

The phylum Cnidaria includes a number of distinct body plans. Some, such as the Scyphozoa, spend most of their lives in the medusa form; (*A*) they are known as sea jellies (formerly called "jellyfish," although they are not related to fish). (*B*) The Hydrozoa are smaller colonial animals with many simple hydra polyps. (*C*) The Anthozoa include the corals, which have small polyps on a hard colony of calcite secreted by the individual animals, and (*D*) sea anemones. (Courtesy of Wikimedia Common)

Figure 12.1 ▲
(continued)

mesoglea, which provide bulk and thickness to their body walls. In a sea jelly, most of their mass is mesoglea, which makes them "jelly-like." They also have a nervous system and muscular system that react to touch, which allows them to sting prey and pull it into their mouth using their tentacles. But they do not have eyes or most other sensory systems, nor do they have a respiratory system or excretory system. They take in all of their oxygen and get rid of all of their waste gases through the surface of their body.

From this basic body arrangement, cnidarians have two fundamental versions (fig. 12.2A): the jelly-like form, or *medusa* (named after the monster in the Greek myth of Perseus and Andromeda, who had snakes for hair and turned anyone to stone who saw her), and the attached anemone-like arrangement, or the *polyp*. In the medusa stage, like sea jellies, the tentacles hang down from the floating body, with the mouth below and the rest of the body above. In the polyp stage, the tentacles are on the top with the mouth, and the body is attached at the bottom.

Although they look very different, both of these stages are actually part of the life cycle of the same organism (fig. 12.2B). Cnidarians reproduced by alternating generations between sexual and asexual reproduction. The polyps are the asexual stage, and they can bud and clone themselves asexually to form huge colonies. Eventually, however, some polyps produce tiny larvae that grow into free-swimming medusae, which have sexual organs. These swarm in the ocean and release sperm and eggs in a coordinated fashion, which fertilize to form a free-swimming larva that eventually settles and grows into a polyp again. Some cnidarians (like *Hydra* in a freshwater pond, or corals) spend almost all of their lives as immobile polyps, having only a brief medusa stage to spread to new habitats and colonize new seafloor. Others, like sea jellies, spend most of their lives in the medusa stage, and only exist as polyps for a brief time.

Most cnidarians, including freshwater hydras, sea anemones, and sea jellies, are soft-bodied and have no hard parts, so they are very rarely fossilized. However, one group of cnidarians—corals—leaves an excellent fossil record. Coral reefs are built of hundreds of polyps that looked like tiny sea anemones, but the base of their ectoderm secretes a large amount of calcite that combines with their neighboring polyps to form a huge coral "skeleton" or coral reef. When we look at a piece of modern coral or a coral fossil, it's hard to imagine this rock was once a living organism, but at one time it was covered with a layer of hundreds of polyps, all busily trapping tiny prey with their tentacles while building their enormous skeletons (see fig. 12.1C).

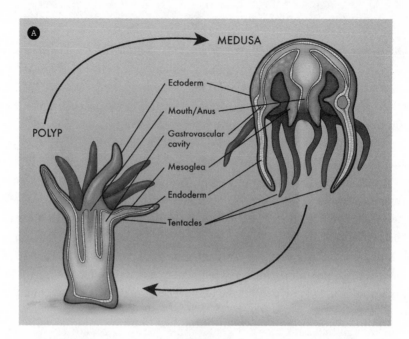

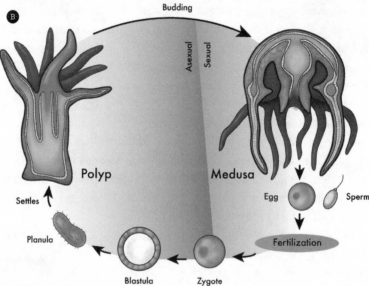

▲

The cnidarians spend their entire lives alternating between two basic arrangements of a hollow body cavity with a mouth opening surrounded by tentacles. (*A*) If they are attached to a surface with their tentacles catching food in the water currents, they are polyps. If they are free-floating with the mouth and tentacles hanging down, they are medusae. (*B*) The typical cnidarian life cycle. All cnidarians alternate generations between the polyp and medusa stage. (Illustration by Mary Persis Williams)

Some corals can live in cold water, or in deep water without much light, but they tend to secrete very small skeletons. The vast majority of reef-building corals have symbiotic algae that live in their tissues. The algae are plants, which provide the coral with oxygen and use up the carbon dioxide they produce. Algae also helps the coral secrete the enormous volume of calcite needed to make their limestone reefs. All plants require light for photosynthesis, so large reef-building corals can only live in very shallow tropical oceans where there is good light penetration and where sand or mud from nearby land does not make the water dark and murky.

Thus coral reefs grow in only a few parts of the world today, mostly in tropical or subtropical shallow oceans far from the influence of nearby rivers and their muds. These areas include the Bahamas, the east coast of Florida (but not most of the Gulf of Mexico side, where the Mississippi mud makes the water dark), and other parts of the Caribbean and Yucatán Peninsula of Mexico; the Persian Gulf; and the South Pacific islands and Great Barrier Reef of Australia; and just a few other places.

Paleontologists believe that ancient corals had similar restrictions, yet we find fossil corals in places like Iowa. How could this be? At various times in geological history, especially during the early Paleozoic, the entire planet was much warmer, sea level was higher, and shallow warm tropical seas drowned the continents. As a result of these factors, fossil corals can be found in many places that are now far from the ocean. Let's take a closer look at three main groups of corals that are commonly fossilized: tabulate corals, horn corals, and modern corals.

Tabulate corals (Order Tabulata). Tabulate corals are built of dense clusters of tiny tubes (*corallites*) all packed together like a box of drinking straws. Each corallite had lots of tiny dividing walls inside it, known as *tabulae* ("little table" in Latin), so this feature gives the group its name. A tiny coral polyp lived in the top of each tube.

Tabulate corals were extremely important as reef builders in the early Paleozoic, especially in the Silurian and Devonian. The "honeycomb coral" (*Favosites*) is a typical Silurian-Devonian coral found in many localities in the Midwest, and it gets its appearance from its "honeycomb" appearance when it is sliced across the top (fig. 12.3A). Another common Silurian reef builder is the "chain coral" (*Halysites*), whose corallites form little loops or chains with gaps between them in top view (fig. 12.3B). Tabulate corals were nearly wiped out during the great Late Devonian extinction event, but they

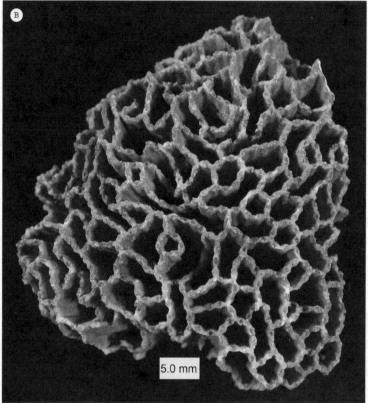

The tabulate corals were built of dense clusters of tube-shaped corallites. The top hole
of the tube would have been occupied by a tiny polyp. (*A*) The tabulate coral *Favosites*,
with the densely packed cluster of corallites giving it a porous appearance on the surface.
(*B*) The chain coral *Halysites*, which had it tubular corallites arranged in chains. (Courtesy of
Wikimedia Commons)

straggled on through the rest of the Paleozoic, only to vanish in the great Permian extinction.

Horn corals (Order Rugosa). The other common group of Paleozoic corals was the horn corals, or rugosids. Many are shaped like the curved horns of a cow, hence their common name. In life, a typical horn coral like *Streptelasma* (fig. 12.4A) or *Zaphrentis* (fig. 12.4B) would have sat on the seafloor with the narrow pointed end embedded in the sediment, and a large anemone-like animal would have lived in the bowl-shaped cavity on the top of the cone, or *calyx*. Most rugosids had a wrinkled outer surface on the "horn," hence their name ("rugose" means "wrinkled" in Latin). The most extreme example of this is the lumpy, irregularly wrinkled Devonian coral known as *Heliophyllum* (fig. 12.4C). Its name means "sun leaf," possibly because, if you look down on the calyx, the septa radiate outward like the rays of the sun.

Horn corals like *Heliophyllum* have proven to be important for other reasons. If you slice one lengthwise and polish it, under the microscope you can see their daily growth bands as well as the annual growth bands caused by the seasons. In the 1960s, paleontologist John Wells counted the number of daily growth bands recorded in a year and found that there were 400 days in a year in the Devonian (about 400 million years ago), not the 365.25 days in a year we know today. In other words, the earth used to spin around its axis much faster, and it has been slowing down ever since. Other studies of different kinds of fossils with daily and annual growth bands confirmed this pattern. This phenomenon was predicted many years ago by astronomers who realized that the earth was being slowed down by the gravitational drag of the moon's tidal pull on the earth. Many million years from now, the earth will finally slow down to a stop, no longer spinning on its axis. This has already happened to the moon, which has been slowed down by the earth's tidal pull of gravity so that one side always faces the earth. The other side of the moon has only been seen by Apollo astronauts who flew around it and photographed it for the first time.

Although most horn corals were solitary, some rugosids clustered together into dense colonies. The most famous of these is *Hexagonaria*, an important reef-building coral of the Devonian (figs. 12.4D–E). When you slice across *Hexagonaria*, you can see that the corallites were packed together tightly to form a hexagonal honeycomb pattern, hence their name. Polished pebbles of *Hexagonaria* are common on the shores of the Great

Figure 12.4 ▲

The rugosids, or "horn corals," lived on the sea bottom with their pointed end embedded in the sediment and an anemone-like animal in the top bowl or calyx. (*A*) The common Ordovician horn coral *Streptelasma*. (*B*) The Silurian-Devonian coral *Zaphrentis*. (*C*) The wrinkled Devonian coral *Heliophyllum*. (*D*) The colonial rugosid *Hexagonaria*. Its corallites are tightly packed together like the cells in a honeycomb, giving it the nickname "honeycomb corals." (*E*) Entire colonies of *Hexagonaria* are often eroded into pebbles, known as "Petoskey stones." ([*A*] Illustration by Mary Persis Williams; [*B–E*] courtesy of Wikimedia Commons)

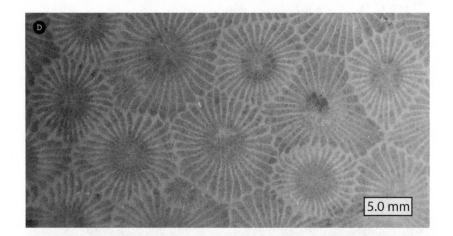

Figure 12.4 ▲
(*continued*)

Lakes and have come to be known as "Petoskey stones." These are the official State Rock of Michigan. Other important late Paleozoic colonial rugosids were *Lithostrotion* and *Syringopora*.

Like tabulates, rugose corals were hit hard by the Late Devonian mass
extinction, which wiped out the immense reefs made of tabulate and corals
and stromatoporoid sponges. The rugosids straggled on through the rest of
the Paleozoic, but they were finally wiped out by the great Permian extinction.

Modern corals (Order Scleractinia, or hexacorals). Both tabulate and rugose
corals vanished in the great Permian extinction, and there were no corals or
reefs of any kind in the Early Triassic. By the Middle Triassic, some group
of soft-bodied anemones evolved the ability to build big calcite skeletons
again, and the modern group of scleractinian corals evolved (see fig. 12.1C).
They are also called hexacorals because the internal dividing walls (septa)
are arranged in multiples of six. (In contrast, the pattern for rugose corals is
multiples of four, hence their old name, "Tetracorallia.")

Scleractinian corals have adopted a wide variety of body shapes, from
the branching corals to massive dense corals like the "brain coral," and
many other shapes. These corals have been evolving rapidly ever since the
Triassic, but they have gone through numerous crises as well when other
reef builders pushed them out of their preferred habitats. For example,
during the Cretaceous coral reefs were almost displaced by huge reefs of
colonial reef-building oysters known as rugosids.

Although it's hard to imagine anything eating stony corals, they have
numerous predators. Parrotfish have hard beaks that are excellent for
crushing hard coral, and sudden population explosions of the crown-of-
thorns sea star destroyed some tracts of the Great Barrier Reef. In addition,
specialized crabs, polychaete worms, snails, and echinoids are adapted to
feeding on coral polyps, and many organisms (clionid sponges, bivalves,
and certain worms) are specialized to bore and burrow into coral skeletons.

Today, one of the great worries of biologists and environmental scientists
is that global warming has been causing the world's oceans to get warmer
and more acidic, driving out their symbiotic algae and thus killing the
world's coral reefs in a process known as *bleaching*. In just the past 30 years,
marine biologists have noted a dramatic worldwide dying off of coral reefs
that were healthy and thriving in the 1980s and 1990s. Many reef biologists
predict that coral reefs will be wiped out all over the planet within a few
decades, causing a mass extinction in the huge number of animals that
depend on coral reefs for their lives.

PHYLUM BRACHIOPODA

Although they are relatively rare today and unfamiliar to most of us, during the Paleozoic the brachiopods, or "lamp shells," were by far the most common of all invertebrate fossils. In many places in the United States, especially in the East and the Midwest, the limestones are chock full of brachiopods. They literally pave the ground (fig. 13.1), and they are excellent index fossils. If you can recognize the basic brachiopod groups, you can easily tell what time it is in the Paleozoic from brachiopods alone. There are only 120 living genera, but there were more than 4,500 fossil genera (900 in the Devonian alone). Nearly all of the modern brachiopods live in hiding places, such as under rocks or burrowed into the mud, to avoid predators. Consequently, brachiopods have long been intensely studied by paleontologists. However, most marine biologists have never seen one, and brachiopods get only a page in a typical invertebrate zoology textbook.

Brachiopods are not familiar to most of us because of their scarcity in modern oceans. The only common name they have is "lamp shells," which was given to one group of brachiopods that resembles the biblical oil lamp. Superficially brachiopods look like clams because their bodies are encased in two hard shells, also called *valves*. However, inside the shell their anatomy is completely different (fig. 13.2A), and they are not closely related to clams or any other kind of mollusc. Even their bivalved shell is very different from the shell of a clam (fig. 13.2B). The best rule of thumb for telling them apart is that brachiopods are symmetrical *through* the valves, so the right half of each shell is the mirror image of the left half. In contrast, a clam

1.0 cm

A dense pavement of brachiopod shells of *Cincinnetina meeki*. Such concentrations of brachiopods are typical in many Paleozoic limestone localities. (Courtesy of Wikimedia Commons)

is a mollusc (an entirely different phylum), and it is symmetrical *between* the valves, so one shell is the mirror image of the other. However, there are a number of oyster-like brachiopods, as well as true oysters and a number of other clams that give up on symmetry altogether, so these rules of thumb don't always apply.

Once you have determined the fossil is a brachiopod shell, the rest of the anatomy is straightforward. One valve is usually larger than the other, and many brachiopods have a small opening on the hinge for the long fleshy stalk known as the *pedicle*, which is used to attach the shell to the substrate (see fig. 13.2A). This has long been known as the *pedicle valve*, although now it is called the *ventral valve* because it usually sits in the ventral (bottom) position. The pedicle gives the phylum its name, Brachiopoda, which means "arm foot" in Greek. Hinged to the ventral valve is the other valve, which originally was called the *brachial valve* but today is called the *dorsal valve*. It is slightly smaller than the ventral valve, and it tends to sit on the top, or dorsal position, in most brachiopods.

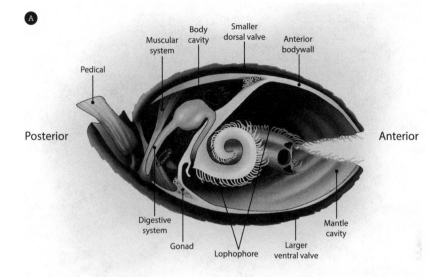

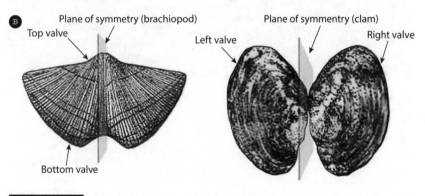

(A) Anatomical features of a typical brachiopod, showing the fleshy pedicle used for attachment, the concentration of organs in the back of the shell near the hinge, and most of the internal shell cavity, which is occupied by the feathery lophophore. (B) The basic symmetry of brachiopods compared to the shells of clams (bivalves). (Illustrations redrawn by Mary Persis Williams)

Inside their paired shells (bivalved shells), brachiopods have a very simple anatomy. Their main feature is a large, feathery, filter-feeding device known as a *lophophore*, which is used to trap microscopic food particles as they pass through the gap in the open shell (fig. 13.3). The food then passes down through the mouth and the digestive tract as it does in most other

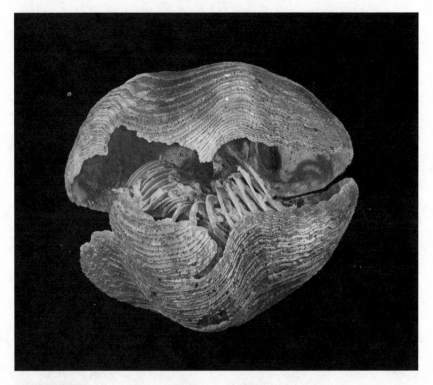

Figure 13.3 ▲

The interior of the brachiopod shell has a lophophore, which is supported by a rigid structure. This example is a spiriferide brachiopod, whose lophophore was shaped like a spiral. (Courtesy of Wikimedia Commons)

coelomates. They have all the other usual organs, such as gonads, excretory system, digestive glands, and a simple circulatory system. Brachiopods have a nervous system, but no eyes. Instead, bristles around the margin of the shell, called *setae*, are sensitive to changes outside the shell and warn the animal to close its shell when danger approaches.

Most important are the muscles that close and open the shell. One pair, known as the adductor muscles, pull the shell closed; a second pair pulls on the lever arm of the internal hinge to pull the shell open. Because they have paired muscles controlling the shell, brachiopods tend to stay closed when they die, and they tend to be buried and fossilized as complete shells. In contrast, clams and other bivalve molluscs have adductor muscles to close the shell, but a flexible ligament in the hinge spring-loads the shell so it opens automatically when the clam is not trying to close the shell. Thus, when clams or scallops die and the adductor muscles relax, they tend to open automatically. The shells often break apart, and it is rare to see both shells preserved together.

Many details of the shell, especially in the hinge area, the shape of the lophophore, and the detailed microscopic structure of the shell, are used in identifying the brachiopods and creating their classification. For a book of this level, however, we will look at only a few anatomical features that are helpful for identifying the external shape of the shell (fig. 13.4). The outside of the shell can have numerous *growth lines*, which form arcs radiating out from the hinge, and fine ribs known as *costae*, which radiate out from the hinge-like spokes. Some shells have large corrugations radiating from the hinge, known as *plications*. Certain groups of brachiopods have a large, trough-like depression on the midline called a *sulcus*, and a corresponding large ridge on the midline of the other shell called a *fold*. The edge of the shells where they join is called the *commissure*. In brachiopods with strongly plications, the commissure has a zigzag shape.

Most brachiopod shells have two shells that bulge outward, so they are *biconvex*. However, some have a convex ventral valve, but a flat dorsal valve, and they are *plano-convex*. Some even have a dorsal valve that sinks down into the body cavity, and they are *concavo-convex*. A few have the same shape but the dorsal valve is convex and the ventral is concave, and they are *convexi-concave*.

BRACHIOPOD IDENTIFICATION

With over 4,500 fossil genera of brachiopods, it is not possible to describe how to identify every possible kind. Instead, we will look at the major orders of brachiopods, which are fairly easy to tell apart, even for the amateur. Knowing which major group of brachiopod you are looking at will enable you to identify the Paleozoic period from which your fossil comes.

INARTICULATE BRACHIOPODS

The most primitive of all the brachiopods are the inarticulates. The name doesn't suggest they can't speak fluently; instead, it refers to the fact that they have no mechanical hinge with teeth and sockets holding the shells together.

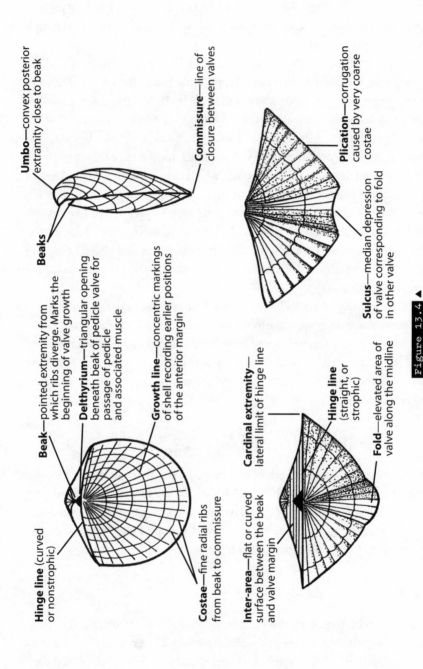

Umbo—convex posterior extremity close to beak

Beaks

Beak—pointed extremity from which ribs diverge. Marks the beginning of valve growth

Delthyrium—triangular opening beneath beak of pedicle valve for passage of pedicle and associated muscle

Growth line—concentric markings of shell recording earlier positions of the anterior margin

Commissure—line of closure between valves

Hinge line (curved or nonstrophic)

Costae—fine radial ribs from beak to commissure

Cardinal extremity—lateral limit of hinge line

Hinge line (straight, or strophic)

Fold—elevated area of valve along the midline

Inter-area—flat or curved surface between the beak and valve margin

Plication—corrugation caused by very coarse costae

Sulcus—median depression of valve corresponding to fold in other valve

Figure 13.4 ▲

Terminology of the external features of the exterior of the brachiopod shell. (Redrawn by Mary Persis Williams)

Figure 13.5 ▲
The living inarticulate brachiopod *Lingula*, shown on the sediment surface after having been extracted from its burrow (see also color figure 3). (Courtesy of Wikimedia Commons).

The hinge is held together only by bands of muscles, with no mechanical articulation. One of the most common brachiopods is alive today and is found in mudflats all over the world. It is called *Lingula* ("little tongue" in Latin) because the shell has a tongue-like shape (fig. 13.5). *Lingula* live buried deep in mudflats, using a very long fleshy stalk-like pedicle to burrow down. This type of brachiopod has been around unchanged since the Cambrian when inarticulates became the first brachiopods on the planet, so they are a living fossil that has persisted for 550 million years. A number of other tongue-shaped, coin-shaped, oval, and disk-shaped inarticulates were found in the Cambrian, but only *Lingula* survives today. Inarticulates are also interesting in that they make their shells out of calcium phosphate (the mineral apatite), the same material used in our bones, rather than the calcite that nearly all other marine invertebrates use.

ARTICULATE BRACHIOPODS

All the rest of the brachiopods are articulate; that is, their shells have a mechanical hinge that holds together even when the animal dies. All of them make their shells out of calcite, the most common building material among marine invertebrates. They are also peculiar in that they have no anus. When the digested food accumulates to their limit, they expel all their waste products out of their digestive tract.

Order Orthida

The orthides are the most primitive of the articulate brachiopods, originating in the Late Cambrian during the heyday of the inarticulates, flourishing in the Ordovician before crashing during the Late Ordovician mass extinction, then straggling on to the end of the Permian extinction. Most are very similar, with a straight hinge and small pedicle opening, and with lots of fine ribs of costae radiating away from the hinge. During the Ordovician, genera like *Orthis, Dinorthis, Hebertella,* and *Resserella* (fig. 13.6A–B) were very common, and they are good index fossils of this period.

Order Strophomenida

The strophomenides are the largest group of articulate brachiopods. They had two great periods of diversification. During the Ordovician, they were by far the most common type of brachiopod, and thus they are an instant indicator of Ordovician rocks. Ordovician strophomenides (fig. 13.7A) tended to look like *Strophomena* or *Rafinesquina*, with long straight hinges giving them a D-shape in top view. These strophomenides also tended to have concavo-convex shells, so when the valves were closed, they were like a pair of bowls nested inside one another and had only a tiny internal volume. They apparently lived with their convex side up, so they arched their shell above the substrate. The lack of a large pedicle opening tells us that they did not attach to anything, but instead they lived on the open seafloor.

The D-shaped strophomenides were nearly wiped out during the Late Ordovician extinction, but the group had another great evolutionary radiation in the Late Paleozoic (Carboniferous-Permian). These brachiopods, known as *productids*, had a cup-shaped ventral valve and a tiny flat or concave dorsal valve that formed a lid on top of the shell (fig. 13.7B–C). They had no pedicle, and they lived on the open seafloor. Most had a dense cluster of spines on their ventral valve that acted as "stilts" or "snowshoes" to prevent their shell from sinking into the sediment. In fossil productids, the delicate spines are usually broken off, but the ventral valve is covered with a dense set of bumps where they once attached.

Productids were by far the most common brachiopods of the Carboniferous and Permian, and they often lived in dense clusters or colonies. In the Permian they evolved some truly strange forms, known as the leptodids.

Figure 13.6 ▲

Typical orthide brachiopods: (A) a slab full of *Orthis*, a typical orthide; (B) the common Ordovician genus *Hebertella*. ([A] Courtesy of Wikimedia Commons; [B] photograph by the author)

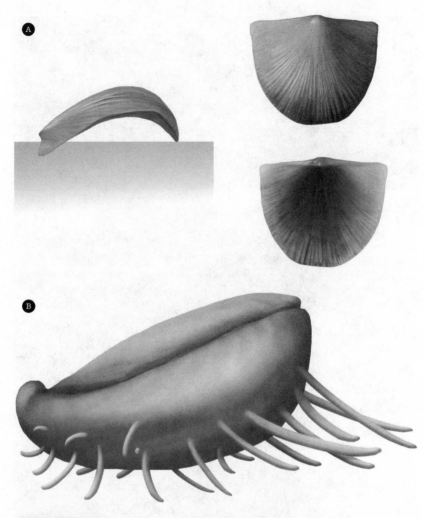

Figure 13.7 ▲

Strophomenide brachiopods: (*A*) typical Ordovician strophomenides, like this *Rafinesquina*, had long straight hinges and looked like the letter "D" in dorsal view, and had concavo-convex shells; (*B*) Late Paleozoic productids were shaped like a cup or dish supported by spines, with a small lid-like dorsal valve to close them; (*C*) ventral view of a Permian productid with the spines still preserved. ([*A*] Illustration by Mary Persis Williams; [*B–C*] courtesy of Wikimedia Commons)

2.0 mm

Figure 13.7 ▲
(*continued*)

These weird creatures had a ventral valve that was shaped like a soap dish that was covered by a dorsal valve that looked like a comb or a grill. In contrast to the flattened leptodids, the other extreme were the richthofenids. These peculiar creatures had a ventral valve that was shaped like an ice-cream cone held up by stilt-like spines, with a tiny lid-like dorsal valve inside the opening of the cone. This was one of many examples of a group evolving into a conical and colonial coral-like shape or oyster-like habitat. A similar shape also happened independently in the rudistid oysters in the Cretaceous.

Order Pentamerida

During the Silurian, the most common group of brachiopods was the pentamerides. They had a robust biconvex shell with a narrow hinge, a large pedicle opening, and a relatively smooth shell covered with fine costae. Their name comes from the fact that their internal shell is subdivided into chambers by a series of small dividing walls. There are five chambers, hence the name "pentamerid" (*penta* is "five" and *meros* is "part" in Greek).

Pentamerides tended to form dense reef-like clusters during the Silurian, often in areas associated with the huge Silurian coral and sponge reefs (fig. 13.8A). In many places in the Midwest, the Silurian rocks are completely replaced by dolomite, including the filling of the shells. However, the shell itself remained preserved in calcite, so when these rocks weather, the shells dissolve away, leaving a steinkern or internal mold of the more resistant dolomite (fig. 13.8B). Their distinctive internal molds have a large cleft in the middle where one of the dividing walls used to be located.

Order Spiriferida

The spirifers are distinguished by their lophophores, which are arranged in a spiral or corkscrew pattern inside the shell (hence their name "spirifer," which means "spire bearing" in Latin). The external shape of *Spirifer, Mucrospirifer, Neospirifer*, and their kin is also distinctive (fig. 13.9A–C), with a long straight hinge, strong plications, and a large fold and sulcus. Their shape resembles a pair of wings. Other spiriferides, like *Atrypa*, were plano-convex with many fine costae and a slightly shorter hinge. Spirifers first appeared in the Ordovician, but during the Devonian they had a huge radiation, and they are a classic index fossil of that period. They were decimated during the Late Devonian extinction, but genera like *Spirifer* flourished in the Mississippian, and the entire group straggled through the rest of the Paleozoic in small numbers. They were nearly wiped out by the great Permian extinction, but they recovered in the Triassic, only to vanish at the end of the Triassic.

All the previous articulate orders are now extinct. Only two articulate groups survived the Mesozoic and are still around today.

Order Rhynchonellida

The rhynchonellides first appeared in the Ordovician, but they persisted through the entire Phanerozoic and are still found in great numbers in certain habitats. Nearly all rhychonellides have a short hinge with a pointed beak and strongly corrugated plications that give them a distinctive zigzag

Figure 13.8 ▲

(A) Pentamerides typically formed dense "reefs" of clustered brachiopods in the Silurian.
(B) Many times, pentamerides are preserved as dolomite internal casts or steinkerns, which
filled their original shell. The small clefts on the steinkerns are where the internal dividing
walls of the brachiopod were located. ([A] Illustration by Mary Persis Williams; [B] courtesy
of Wikimedia Commons)

Figure 13.9 ▲

Spirifers were the most common brachiopods in the Devonian and Mississippian. They all had long straight hinges, a deep fold and sulcus, and a spiral lophophore inside. (*A*) *Mucrospirifer* was a very long-hinged form whose shape resembled a pair of wings. It is an index fossil of the Devonian. A common Mississippian brachiopod is *Spirifer*: (*B*) dorsal view, with the prominent fold on the midline; (*C*) ventral view, with a deep sulcus on the midline. (Photographs by the author)

commissure (fig.13.10A). This body form was very conservative and typifies nearly all of their order, so they are easy to recognize.

Order Terebratulida

The other surviving group of articulate brachiopods is the terebratulides. These are the most common form alive today, and they are shaped like the

Figure 13.10 ▲

(*A*) Rhychonellides typically have a sharp pointed hinge and short hinge line, and a corrugated shell with deep plications, producing a zigzag commissure. Typical of them is *Rhynchotrema*. (*B*) The common living brachiopods are terebratulides, with a smooth shell shaped like a biblical oil lamp. This is *Terebratula maugeri*. (Courtesy of Wikimedia Commons)

biblical oil lamp, hence the name "lamp shells" (fig. 13.10B). They are strongly biconvex, with a narrow hinge. They have a large pedicle, and the beak of the ventral valve curves around with a large pedicle opening. Living terebratulides attach to a hard surface and can orient their shells in any position to take advantage of currents.

PHYLUM BRYOZOA

The other living phylum of animals that have lophophores are the bryozoans, or "moss animals." In contrast to the relatively large brachiopods, bryozoans were tiny creatures similar to the polyps of corals. Like corals, they lived colonially on a massive skeleton of calcite secreted by all of them collectively, and they filter-fed on small food particles from the seawater that passed through their lophophore (fig. 14.1). However, individual bryozoan animals are smaller than even coral polyps. If you are looking at a piece of a colonial fossil and trying to decide if it's coral or bryozoan, the rule of thumb is this: bryozoans leave pinprick-sized holes, whereas the openings for coral polyps tend to be larger.

Bryozoans are still very common in the world's oceans today, with more than 3,500 living species, and 15,000 or more fossil species as well. They are sometimes known as moss animals because a living bryozoan colony looks like it is covered by a fuzzy coat of moss when all the individual animals are reaching out of their tiny holes and feeding. Despite their incredible living diversity, beachcombers and marine biologists seldom notice them; they are very tiny and are often mistaken for moss or algae.

However, bryozoans are most definitely not built like coral polyps. Instead, they have a large, feathery lophophore that protrudes out of their little hole in the colony when they are feeding. At the base of the lophophore is a U-shaped digestive tract, which takes food brought down from the lophophore and processes it, then excretes it back out the other end through the anus. The individual animals also have simple gonads, excretory

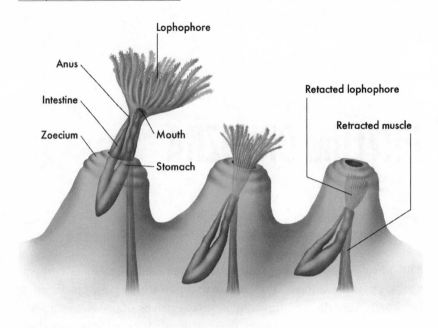

Lophophore

Anus

Intestine

Zoecium

Mouth

Stomach

Retacted lophophore

Retracted muscle

Figure 14.1 ▲

Anatomy of a bryozoan colony, showing the basic structure of the individual animals. (Illustration by Mary Persis Williams)

systems, and a set of retractor muscles that pull them back into their holes. Most of them have a small lid that closes behind them when they retract, sealing them in their chambers when they are exposed to danger or drying conditions.

Sadly, it is difficult for nonspecialists to identify most bryozoans because their features are extremely tiny, and microscopes and thin sections are required to study them properly. For a book at this level, I won't try to go into that level of detail. Instead, I will discuss just a few of the more distinctive Paleozoic colonial forms.

MASSIVE AND BRANCHING BRYOZOANS

In the early Paleozoic (especially the Ordovician), there was a great radiation of two groups, the cryptostomes and trepostomes. They are difficult to tell apart without a microscope, so I will just note that they came in a wide

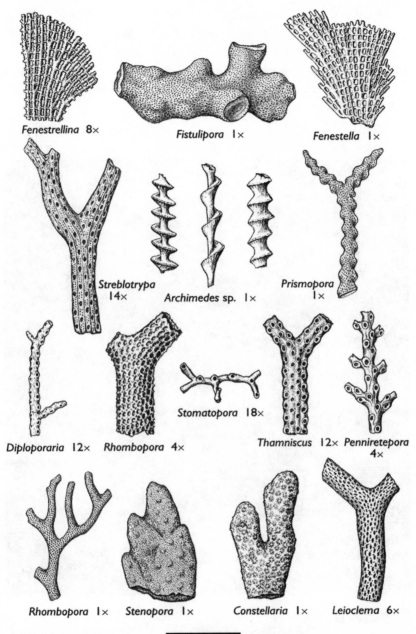

Fenestrellina 8×

Fistulipora 1×

Fenestella 1×

Streblotrypa 14×

Archimedes sp. 1×

Prismopora 1×

Diploporaria 12×

Rhombopora 4×

Stomatopora 18×

Thamniscus 12×

Penniretepora 4×

Rhombopora 1×

Stenopora 1×

Constellaria 1×

Leioclema 6×

Figure 14.2 ▲

Sketches of common fossil bryozoans. (Courtesy of Wikimedia Commons)

variety of body forms. Some, like *Prasopora* (fig. 14.3A), *Rhombopora*, and *Constellaria* (fig. 14.3B), formed massive disk-like colonies that occasionally reached 2 feet across, but most were usually only a few inches long. Others, like *Dekayella, Prismopora, Streblotrypa, Leioclema*, and *Thamniscus*, grew into distinctive branching forms, with each branch covered with hundreds of tiny pinprick-sized holes (see fig. 14.2).

1.0 cm

Figure 14.3 ▲

Typical bryozoans: (*A*) the massive bryozoan *Constellaria*, which gets its name from the star patterns on its surface; (*B*) the lumpy bryozoan *Prasopora*; (*C*) the lacy bryozoan *Archimedes* with lacework arranged in a corkscrew spiral around a central stem, which is what usually fossilizes; and (*D*; color figure 4) restoration of *Archimedes* in life (see also color plate 2). ([*A–C*] Courtesy of Wikimedia Commons; [*D*] illustration by Mary Persis Williams)

3.0 mm

Figure 14.3 ▲
(continued)

Figure 14.3 ▲
(*continued*)

In the late Paleozoic (especially the Mississippian), there was a great radiation of another group, known as the fenestrate or *fenestrellid* bryozoans, such as *Fenestella* and *Fenestrellina* (see fig. 14.2). These formed a lacy framework with a grill-like pattern in detail. Their name refers to the window-like openings in the grillwork (*fenestra* is "window" in Latin). They are very common in Mississippian limestones and shales.

The most distinctive of all the fenestrellids was a genus named *Archimedes*. This bryozoan colony was built like a large corkscrew in shape, with the lacy fan of grillwork fanning around the spiral (fig. 14.3C–D). Their tiny corkscrew-like central columns are extremely common in some Mississippian shales, and they are a good index fossil of the Mississippian. They were named after the famous Hellenistic Greek inventor, scientist, and mathematician Archimedes. One of his many famous inventions was a water pump known as the "Archimedes screw." It was built of a long tube with a corkscrew-like set of blades inside. If you plunged one end of the tube into water and turned the corkscrew blades, they lifted water up the tube to a different level. The paleontologist who first named this fossil was inspired by the corkscrew-like shape of the fossil and named it after the famous Greek inventor of that device.

The trepostome, cryptostome, and fenestrate bryozoans dominated the Paleozoic, then all of their huge diversity was wiped out by the great Permian mass extinction, along with most of the brachiopods, the tabulate and rugose corals, and most other common Paleozoic groups. Only a few Paleozoic groups, such as the cyclostomes, managed to survive. During the Mesozoic, bryozoans slowly recovered, but they were dominated by a Jurassic group known as the cheilostomes, which are still the most diverse type of bryozoans in modern oceans. Today, most bryozoans are found encrusting shells of other animals or living in small clumps and clusters on the seafloor.

PHYLUM ARTHROPODA

The "jointed legged" animals of the phylum Arthropoda are by far the most numerous, diverse, and successful creatures on this planet. The arthropods include not only insects but also spiders, scorpions, millipedes, centipedes, crustaceans, the extinct trilobites, and dozens of other groups (fig. 15.1). There are over a million different species of arthropods, and some biologists think there may be many millions more, most having not yet been described. Insects alone make up 870,000 species of which there are 340,000 species of beetles. When the great biologist J. B. S. Haldane was asked what nature told him about the Divine, he reportedly said that "God has an inordinate fondness for beetles." In contrast, there are only 42,000 species of backboned animals and only about 4,000 species of mammals. By nearly any reckoning, arthropods make up 95 to 99 percent of the animal species on Earth (fig. 15.2).

In addition to their diversity, arthropods are also the most numerous animals on the planet. A typical anthill or termite nest might have thousands of individuals. A single pair of cockroaches, if they have unlimited resources and space, could have 164 billion offspring in seven months. In the tropical rainforests, a single hectare may only contain a handful of birds and mammals, but it will have over a billion arthropods, especially mites, bees, ants, termites, wasps, moths, and flies. In the ocean, the plankton in a single cubic meter of seawater may include almost a million tiny crustaceans, especially copepods, krill, ostracodes, and other shrimp relatives.

Arthropods can live in almost any habitat on the planet, from the oceans to fresh water to the land, and many can fly through the skies. They can

Figure 15.1 ▲

Arthropods come in a wide variety of body forms, including such extinct groups as (*top row*) trilobites and eurypterids, (*second row*) chelicerates like scorpions as well as crustaceans, and (*bottom row*) myriapods like centipedes, plus the largest group of all, the insects. (Courtesy of Wikimedia Commons)

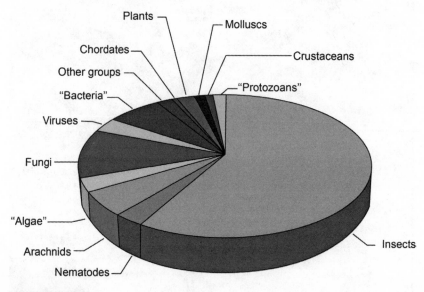

Figure 15.2 ▲

This pie chart shows the diversity of species of the various organisms. Arthropods include arachnids and crustaceans as well as insects, but insects alone outnumber all other groups on the planet combined. In contrast, we chordates (fish, amphibians, reptiles, birds, and mammals) are only a tiny part of the diversity of life. (Courtesy of Wikimedia Commons)

survive in the hottest deserts, on the coldest ice, and in the saltiest lagoon. Many are parasites that live inside or outside other organisms.

What makes arthropods so successful? Certainly, the fact that they can reproduce rapidly in the right conditions helps. They can lay hundreds of eggs that hatch out and develop, and then they can breed again in a matter of days or weeks. Their small body size means they can live in dense numbers on a limited resource, and they can subdivide into many small subniches that are very closely packed. Most important of all, however, is their modular body plan, which consists of a series of segments that can be added or subtracted or modified. On each segment is at least one pair of jointed appendages, which give the phylum its name: *arthros* means "joint" and *podos* means "foot" in Greek. These appendages can be modified into legs, mouthparts, antennae, claws, pincers, swimming paddles, copulatory structures, and a great variety of other possible functions. This modular construction gives them great evolutionary flexibility. Just by changing their embryonic development, they can end up with more or fewer segments, and limbs can be modified as they grow. Some groups (such as the metamorphosis of a caterpillar into a butterfly) can completely change their fundamental body plan in just a few weeks by rearranging and modifying the modules that control each segment.

Another factor that helps them is their external hard shell or skeleton, known as an *exoskeleton*. In most groups, it is made entirely of the carbohydrate *chitin*, but in others (like trilobites) it is reinforced with the mineral calcite. The exoskeleton provides support for the body, making it a hollow tube or shell, and the muscles run within it to make it move. It also provides some protection against predators because it is harder than the soft bodies of worms or many other invertebrates, and the exoskeleton also protects against drying out. This allowed arthropods (first millipedes, then centipedes, scorpions, and finally insects like silverfish and springtails) to emerge onto land in the Late Ordovician, and they continued to flourish for more than 100 million years before the fish-like amphibians managed to crawl onto land.

About the only ecological niche that arthropods don't dominate is the niche for large body size. As animals get larger, their volume (and thus their mass) increases by a cube of their linear dimensions, so organisms get heavier much quicker than they get longer. Their length may increase by only a few centimeters, but their volume increases so fast that they soon test the limits of supporting the weight of their bodies under the rapid increase in gravity. Arthropods have another key limitation: their exoskeleton cannot grow with them, so they must *molt* by breaking out of it and hardening a new exoskeleton. This is not a problem when they are tiny, but as they get larger and larger, the effects of gravity increase dramatically. Once they pass a certain size, their unsupported soft body after a molt would collapse without support. So the house-sized ants and praying mantises from old-fashioned science fiction movies are a complete impossibility.

The largest land arthropods (fig. 15.3) today are goliath beetles, which are at most about 4.3 inches (110 millimeters) long. However, during the Carboniferous, gigantic millipedes like *Arthropleura* were over 10 feet (3 meters) long, dragonflies had wingspans of almost 2.5 feet (70 centimeters) across, and cockroaches were almost 1 foot (30 centimeters) long. The Carboniferous was a time when the atmosphere was extremely rich in oxygen, which allowed arthropods (which have relatively inefficient

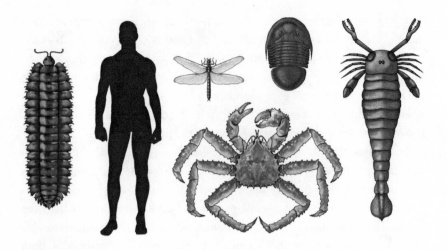

Figure 15.3 ▲

A few examples of giant arthropods. On the left is the huge millipede relative *Arthropleura* from the Carboniferous coal swamps. On the extreme right is the eurypterid (sea scorpion) *Pterygotus*. In the middle at the bottom is the king crab, the largest living arthropod. Above it on the left is the eagle-sized Carboniferous dragonfly *Meganeura*, and on the right the largest of the trilobites, *Isotelus rex*. (Illustration by Mary Persis Williams)

respiratory systems) to get bigger than they ever had been at any other time. The higher density of water supports their bodies better than air, so larger-bodied arthropods exist in the water. The largest today is the king crab, whose legs can span 10 feet (3 meters), but in the past giant eurypterids (sea scorpions) were over 8 feet (2 meters) long.

Even though arthropods are the most successful, diverse, and abundant animals on the planet, they have a relatively poor fossil record. Only about 30,000 species have been described from fossils, despite the fact that millions are alive now and that many more millions must have lived in the geologic past. Most arthropods have a chitinous exoskeleton, which only fossilizes in exceptional conditions. The one group that fossilizes well is the trilobites, which reinforced their chitinous sheath with the mineral calcite, so there are about 2,000 genera and thousands of species of trilobites. Most collectors or paleontologists will not have the opportunity to collect fossil insects or crustaceans or other types of arthropods. So I will focus instead on the trilobites and a few other extraordinary extinct groups from the geological past.

SUBPHYLUM TRILOBITOMORPHA: TRILOBITES

Trilobites are the most popular fossil among collectors, as well as among many professional paleontologists. Not only are they abundant and easy to collect in many places, but they have an extraordinary variety of shapes, including some with weird appendages and spines, and even some with amazing eyes (fig. 15.4). Trilobites have fascinated people for a long time. A Silurian trilobite carved into an amulet was found in a 15,000-year-old Paleolithic rock shelter at Arcy-sur-Cure, France. Australian aborigines chipped a Cambrian trilobite preserved in chert to form an implement. Both of these specimens were clearly imported from a long distance because they were not present in the local area. The Ute people used to make amulets out of the common Middle Cambrian trilobite *Elrathia kingi*, from the House Range of western Utah. They called it *timpe khanitza pachavee*, or "little water bug in stone house." This fossil is so abundant that it can be commercially mined with backhoes; it is found in virtually every rock shop and commercial fossil seller's catalog.

Trilobites first appeared in the third stage of the Cambrian (Atdabanian Stage), and they are by far the most common fossils of the Cambrian, both because they diversified rapidly and because most other Cambrian animals were soft bodied and did not have hard, fossilizable body parts. By the Late Cambrian, there were 65 families with over 300 genera. In Cambrian shales or limestones, virtually every outcrop yields some trilobites, and they evolved so rapidly that we tell time in the Cambrian with them.

By the Ordovician, trilobites began to decline in importance as other invertebrate groups diversified and came to dominate the seafloor. In addition, trilobites faced their first large predators in the enormous shelled nautiloids (see chapter 16). The groups that survived became much more specialized and distinctive, with adaptations for rolling up for protection, burrowing, swimming, or hiding from their predators. Trilobites were hit hard during the Late Ordovician extinction event. During the Silurian and Devonian, there were only about 60 genera in 17 families. The Late Devonian extinction nearly wiped trilobites out, leaving only a handful of genera in a single family that survived until the end of the Paleozoic. They finally vanished during the great Permian extinction event that wiped out tabulate and rugose corals, most groups of brachiopods and bryozoans, and many other Paleozoic groups.

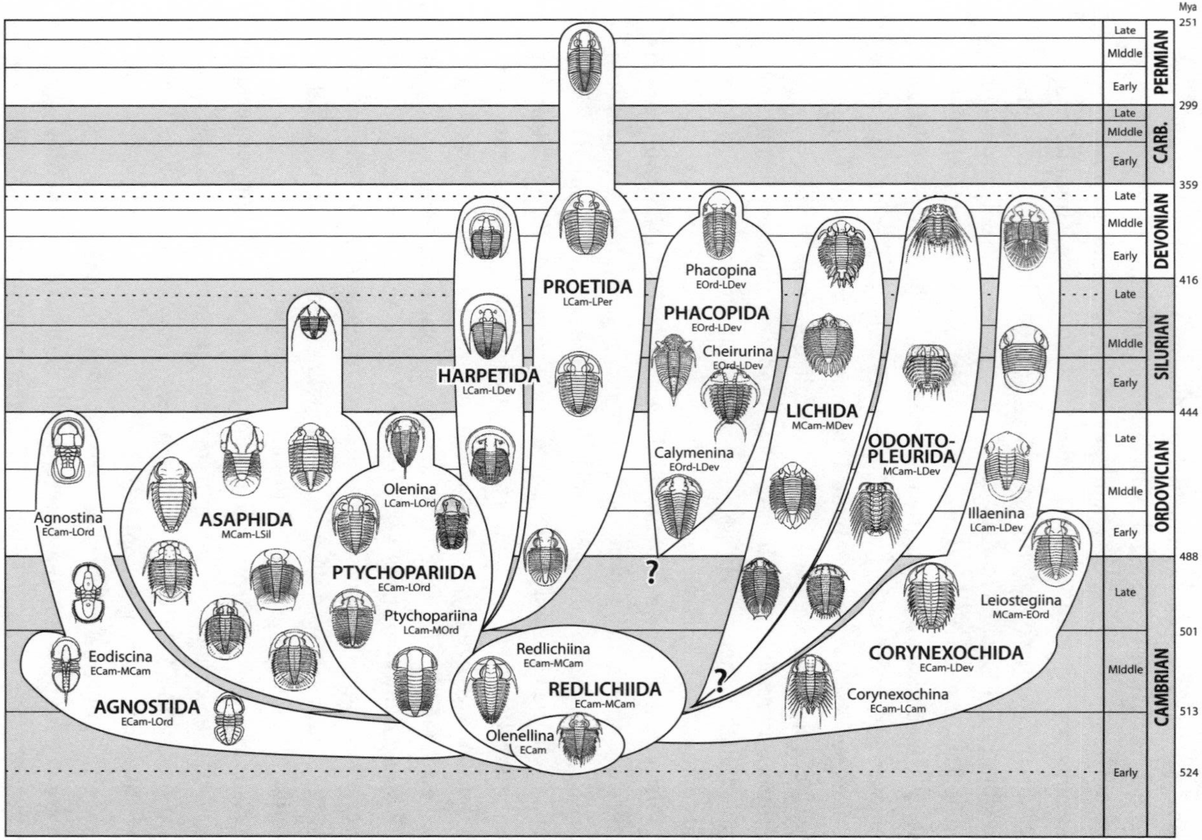

Figure 15.4 ▲

Diagram showing trilobite diversification. (Redrawn from several sources)

The body of the trilobites can be divided into three parts (fig. 15.5A): the head shield (*cephalon*), the middle part or *thorax* made of many segments, and the tail shield, or *pygidium*. In addition, the trilobite shell has three lobes (hence the name "trilobite") from side to side: a central *axial lobe*, and the two sides or *pleural lobes*. The axial lobe on the cephalon forms a bulbous ridge called the *glabella*. Each side of the cephalon adjacent to the glabella is called the "cheeks," and they are split by a line called the *facial suture*. The pattern of the suture is valuable in identifying major group of trilobites. In primitive forms, the back end of the suture splits the back of the cephalon (known as *opisthoparian* suture), but in one specialized group with a *proparian* suture, the back end of the suture splits the side of the cephalon in front of the *genal angle* (the outer back corner of the cephalon). When the trilobite molted, it split the cephalon along this suture. The part that remained attached to the glabella is called the *fixed cheek*, and the two parts that split away are called the *free cheeks*. Many trilobite fossils consist only of a partial cephalon: the glabella plus fixed cheeks (known as the *cranidium*). However, for many trilobites, these are often enough to identify the species.

The eyes are usually on each side of the glabella, right along the facial suture. They may be simple semicircular eyes with hundreds of tiny closely packed calcite lenses (*holochroal* eye) or large elevated bulging eyes with a handful of lenses, each built of two nested parts that correct for spherical aberration (*schizochroal* eyes). A few trilobites had no eyes at all, and they must have lived in dark muddy bottoms where there was no need for vision.

The cephalon has the most diagnostic features, but in complete trilobites, the number and shape of the thoracic segments, and the presence of spines in certain segments, is often diagnostic. The earliest trilobites had no fusion of the tail segments into a pygidium at all, but once trilobites evolved the pygidium, it took lots of shapes and often diagnoses certain trilobite groups.

Trilobites have an exoskeleton made out of calcified chitin, so the external shell preserves quite readily, even though they often break apart during molts. In fact, most fossils of trilobites consist of only one or two segments because they are simply shed molts; the animal probably swam away to live another day. Trilobite specialists usually need only a few diagnostic parts, such as the cranidium, to identify most species.

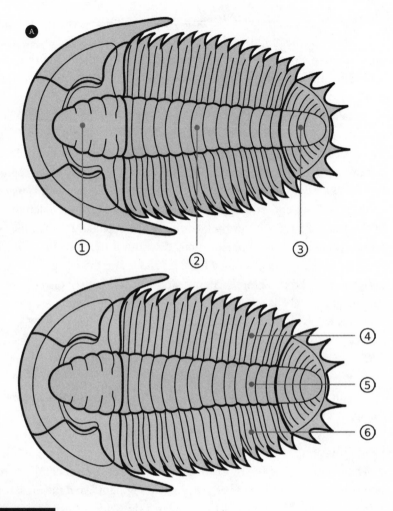

Figure 15.5 ▲

(*A*) The anatomy of the trilobite shell: The three parts of the shell from front to back are
(1) the cephalon or "head shield," (2) the thorax, and (3) the pygidium or "tail shield." Number
(1) also points to the "nose" or glabella. On each side of the glabella are the eyes. The two
bold lines running through the eyes are the facial sutures, lines along which the cephalon
splits during molting. The two areas outside the facial sutures are the "free cheeks"; the
segment of the cephalon between them including the glabella is called the "fixed cheeks."
The three "lobes" of the trilobite from side to side are (5) the axial lobe and (4, 6) the
two pleural lobes. (*B*; color figure 5) Occasionally, black shales preserve the soft parts
of trilobites, here replaced by pyrite. This specimen shows the legs, gills, and antennae.
(Courtesy of Wikimedia Commons)

Figure 15.5 ▲
(continued)

Although only the hard calcified shells of trilobites are usually found today, like all arthropods trilobites had legs, antennae, gills, and other appendages surrounded only by their exoskeleton of chitin but uncalcified. Under the right circumstances these softer parts may be preserved, such as in deepwater black shales where pyrite replaces the organic tissues (fig. 15.5B). This provides a rare glimpse at what the legs, gills, and antennae of other trilobites must have looked like.

With about 2,000 genera of trilobites, it is not possible to talk about every group in this book. The sections below summarize some of the most distinctive and commonly collected genera and orders.

ORDER REDLICHIIDA

The redlichiids are the most primitive group of trilobites, starting in the late Early Cambrian and vanishing at the end of the Middle Cambrian. The most familiar family is the extremely primitive Olenellidae (fig. 15.6A), which are collected in many places in the western United States.

Olenellids were so primitive that they did not fuse their final segments into a pygidium but had a large spike on their tail instead. They had simple half-moon shaped eyes on each side of a glabella with a round knob at the front end, and large genal spines on the outer corners of the cephalon. They had lots of thoracic segments, often with large spines on the "shoulders" of the first few thoracic segments. Early forms like Olenellidae had no visible suture, but later redlichiids such as *Paradoxides* had opisthoparian sutures (fig. 15.6B–C).

ORDER PTYCHOPARIIDA

The ptychopariids is a large wastebasket group for many different primitive trilobites that dominated the Middle and Late Cambrian. They had a simple glabella that tapers toward the front, a large area in front of the glabella, and many thoracic segments.

The classically well-known Cambrian trilobites include *Ptychoparia, Olenus, Triarthrus,* and the most common trilobite on the commercial market, *Elrathia kingi* (fig. 15.7A). This last species is mined in huge quantities from the Middle Cambrian Wheeler Shale in the House Range of Utah and is sold in every commercial market, website, and rock shop in the world.

Examples of very primitive trilobites from the Early-Middle Cambrian, often put in the "wastebasket" group Redlichiida. (*A*) The primitive earliest Cambrian trilobite *Olenellus*, with its simple eyes, many thoracic segments with spines, and a tail spike (telson) instead of a pygidium. (*B*) *Paradoxides*, an important Middle Cambrian trilobite. (*C*) A fossil of *Paradoxides davidis*, the largest trilobite of the Cambrian, which reached 15 inches (37 centimeters) in length. ([*A*–*B*] Illustrations by Mary Persis Williams; [*C*] courtesy of Wikimedia Commons)

Figure 15.7 ▲

(A) The common Middle Cambrian ptychopariid trilobite, *Elrathia kingi*. (B) *Lancastria elongata*, a corynexochid trilobite. (C) *Peronopsis interstrictus*, a tiny agnostid trilobite, with no eyes, a large cephalon and pygidium, but only a few thoracic segments. ([A, C] Illustrations by Mary Persis Williams; [B] courtesy of Wikimedia Commons)

ORDER CORYNEXOCHIDA

The corynexochids are another important Early to Middle Cambrian group. They have a box-like glabella and subparallel facial sutures that are opist-hoparian (fig. 15.7B). The thorax has only seven or eight segments, and some have a large pygidium that is roughly the same shape and size as the cephalon (*isopygous*).

ORDER AGNOSTIDA

These tiny peculiar fossils are so unusual that not all scientists agree that they are trilobites; they may just be a related group of arthropods (fig. 15.7C). They have a button-shaped pygidium the same size and shape as the cephalon (isopygous), so it is the eyes that tell you which is front and which is back. Only two or three short thoracic segments separate their cephalon and pygidium. They are tiny (only a few millimeters long) and may have lived in the plankton rather than grubbing in the bottom muds as most trilobites did. They are most common in the Early to Middle Cambrian but straggled on into the Ordovician. Huge numbers of the tiny agnostid *Peronopsis interstrictus* are collected from the Wheeler Shale in the House Range of Utah, and they can be found in collections and rock shops everywhere.

ORDERS ASAPHINA AND ILLAENIDA

During the Ordovician, most of the unspecialized Cambrian groups declined or vanished, replaced by many highly distinctive and specialized groups. This was probably due to the pressure of new predators, especially the huge new nautiloids, which could see and grab any trilobite within reach of their tentacles (chapter 16). Two groups, the asaphids and illaenids, coped with this by becoming "snowplow" burrowers and living just beneath the sea-floor surface sediments to hide from predators. In both orders, they have a large smooth cephalon shaped like a plow and an isopygous pygidium that mirrored the shape of the cephalon except for the eyes (fig. 15.8A–C). They had only six to nine thoracic segments. Typical asaphids include *Homotelus* (fig. 15.8C) and *Isotelus rex* (fig. 15.8A–B), the largest trilobite known, which reached almost a foot in length. Illaenids like *Bumastus* and *Illaenus* are also common in collections and on the commercial market.

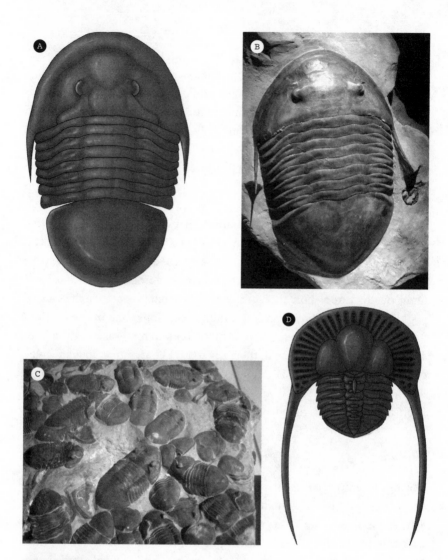

Figure 15.8 ▲

(A) Asaphids are represented by the "snowplow trilobite," *Isotelus*. (B) A specimen of the giant trilobite *Isotelus rex* from the Ordovician of the Cincinnati area. (C) Another common asaphid is *Homotelus*, and death assemblages like this one from the Ordovician can be found in the Criner Hills of Oklahoma. (D) Trinucleid trilobites are typified by *Cryptolithus*, the "lace collar trilobite," with its distinctive "lacy" brim on the cephalon, prominent glabella, enormous cheek spines, and tiny thorax and pygidium. ([A, D] Illustrations by Mary Persis Williams; [B–C] courtesy of Wikimedia Commons)

ORDER TRINUCLEIDA

Another distinctive group of mainly Ordovician trilobites were the trinucleids. Amateur collectors often call them "lace collar trilobites" because they have a huge cephalon and tiny thorax-pygidium, with a perforated "lacy" edge around the brim of the cephalon. They also have a large knob-like glabella that looks like a big nose. Trinucleids like *Cryptolithus* (fig. 15.8D) and *Trinucleus* are common Ordovician fossils and probably survived due to their tiny body size and by burrowing just beneath the surface of the seafloor.

ORDER HARPIDA

Harpids are highly distinctive trilobites, recognized by their broad horseshoe-shaped cephalic brim that wraps around the small thorax and pygidium (fig. 15.9A). The broad cephalic brim was probably suited for plowing through the seafloor sediment.

Unlike trinucleids, harpids had small eyes on tubercles, a glabella that tapers to the front, and lots of thoracic segments.

ORDER LICHIDA

The lichids are another peculiar and distinctive group of trilobites. Their glabella stretches to the front edge of the cephalon and is subdivided by long glabellar furrows, and they had relatively small free cheeks. Many of them, like *Arctinurus* (fig. 15.9B) and *Radiaspis*, had glabella that taper sharply forward with a small "prow" on the front. Their most distinctive feature is their relatively large pygidium, which was often larger than the cephalon and was made of three pairs of expanded spiny pleural segments.

ORDER ODONTOPLEURIDA

Odontopleurids are easy to recognize because most of them are extremely spiny, with spines sticking up from their cephalic brim, genal angle, top of the glabella, all along the thorax, and even from the back of the pygidium (fig. 15.9C). They were relatively rare but were most diverse in the Devonian. They are collected from the Devonian rocks of the Tindouf Basin of Morocco and can be found in many commercial markets.

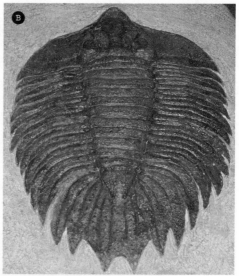
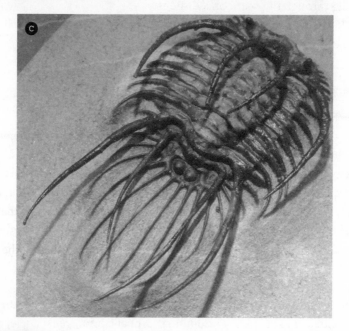

ORDER PHACOPIDA

One of the most interesting and distinctive of the trilobite orders was the phacopids. They are the only group with proparian facial sutures, and most of them were capable of rolling up into a tight ball like sowbugs or pillbugs or "roly-poly bugs" (which are not true bugs but isopod crustaceans that have moved from the tide pools to the land) found in leaf litter today. In the Ordovician and Silurian, they were represented by the distinctive calymenids such as *Calymene* and *Flexicalymene* (fig. 15.10A–B). Calymenids had lots of thoracic segments, and they even had segmented furrows on the glabella, with their eyes out on stalks. Another branch includes the Silurian index fossil *Dalmanites* (fig. 15.10C), and the common Devonian phacopids, including *Phacops, Greenops, Eldredgeops* (fig. 15.10D–E), and others. These phacopids have a broad curved cephalon and a broad glabella that expands forward covered with tiny bumps. They had large schizochroal eyes that stick up above the cephalon, giving them almost a spherical view of the world above them, with stereovision between the lenses. Some, like *Walliserops* (fig. 15.10F) have spines not only along the top and above their eyes, but also a strange trident-like spine in front.

ORDER PROETIDA

The Late Devonian extinction event wiped out the phacopids and odon-topleurids and the other dominant Devonian trilobite groups; only one order managed to survive through the rest of the Paleozoic. Proetids were small trilobites with relatively primitive body forms, a large vaulted gla-bella, opisthoparian sutures, and most of them are isopygous. They had primitive holochroal eyes and a furrowed pygidium without spines. When these last stragglers of the trilobite radiation finally vanished in the great Permian mass extinction, trilobites were completely extinct.

SUBPHYLUM CHELICERATA

The chelicerates include the arachnids (spiders, scorpions, whip scorpions, pseudoscorpions, harvestmen or "daddy long legs," mites, ticks, and chiggers) plus the merostomes. With over 70,000 living and extinct species, the chelicerates are the second largest group of arthropods after insects.

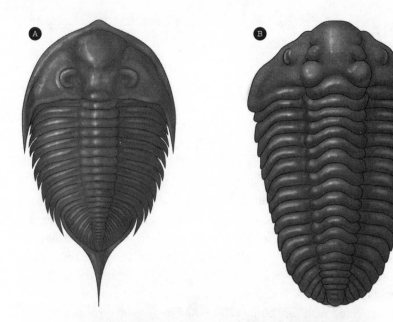

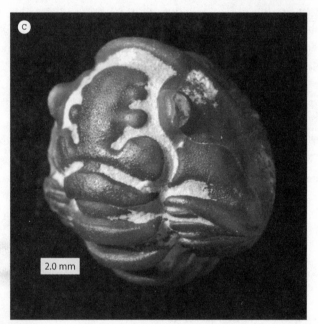

(A) The Silurian index fossil *Dalmanites*, with the pointed spiky pygidium and large schizochroal eyes. (B) *Flexicalymene*, a typical Ordovician calymenid, with lots of segments in the thorax, eyes on stalks, and a furrowed glabella. (C) *Flexicalymene* rolled up in a defensive ball. (D–E) The distinctive Devonian phacopid trilobite *Phacops*, with huge schizochroal eyes, a broad cephalon covered in bumps, and many thoracic segments. (F) *Walliserops*, a spiny phacopid with a remarkable spiny fork on the front and hooks on its eyes. ([A–B, E] Illustrations by Mary Persis Williams; [C–D, F] courtesy of Wikimedia Commons)

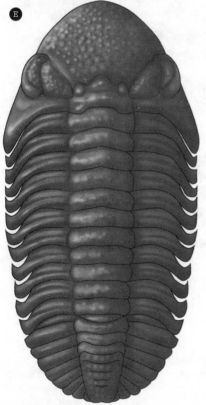

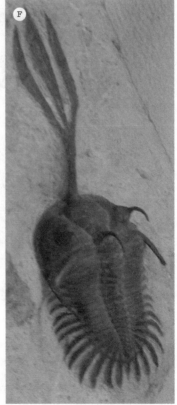

Figure 15.10 ▲
(continued)

They all have a distinctive anatomy of head region (prosoma or cephalo-thorax) and abdomen (or opisthoma). They do not have antennae; instead, their first pair of appendages, known as chelicerae, are found in their mouthparts. Their second pair of appendages, the pedipalps, sometimes serve as mouthparts, but in scorpions, some spiders, and eurypterids, they are modified into pincers for capturing prey.

The most familiar and only living merostomes are the horseshoe crabs (fig. 15.11A), which are not closely related to true crabs (Subphylum Crustacea) at all. They are often called living fossils because some of the Mesozoic forms (fig. 15.11B) look somewhat like modern species, but in the Paleozoic they experimented with many other body shapes as well.

Figure 15.11 ▲

Merostomes: (A) The living horseshoe crab, *Limulus polyphemus*, still common on the Atlantic Coast during very high tides; (B) a fossil horseshoe crab, *Mesolimulus walchi*, from the Upper Jurassic Solnhofen Limestone of Germany. (C) Eurypterids came in many different sizes and shapes. (D) Specimens of smaller *Eurypterus remipes*, a common fossil from the Bertie Limestone of New York. (E) Life-sized model of a huge eurypterid. ([A–D] Courtesy of Wikimedia Commons; [E] photograph by the author)

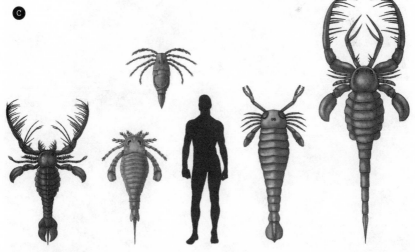

Figure 15.11 ▲
(continued)

Figure 15.11 ▲

(*continued*)

For paleontologists, however, the most impressive chelicerates in the fossil record are the "sea scorpions," or *eurypterids*. Their nickname is a misnomer because they are not closely related to scorpions (which are arachnids, not merostomes) and apparently lived in fresh water and not just in the sea. However, their body form is reminiscent of scorpions with large pincers out front, and spidery legs flaring out from their prosoma (fig. 15.11C). More impressive is the fact that they were one of the world's largest predators during the Silurian. The largest nearly complete euryp-teid fossil is *Pterygotus* (fig. 15.11D), which as about 7 feet (2.1 meters) long, and the giant *Jaekelopterus*, which may have been about 8.2 feet (2.5 meters) long. *Pterygotus* had a long body with a flat swimming tail and a pair of appendages modified into paddles, with huge pincers up front. *Stylonurus* was more spider-like with a spike on its tail and no swimming paddles. *Mixopterus* had long spiny chelicherae and pedipalps, spiny walking legs, and a tail like a scorpion.

Altogether, there were about 70 genera and 250 species of eurypterids, ranging from the Ordovician to the Permian. They were the largest marine predators of the Silurian, and they continued that role in the Devonian. In the United States, the famous Silurian Bertie Limestone of upstate New York produces a majority of the specimens of *Eurypterus remipes* (fig. 15.11E) found in collections and in the commercial market.

PHYLUM MOLLUSCA

Next to the arthropods, the molluscs are the most abundant and success-ful phylum of animals on Earth, with over 130,000 living species of snails, slugs, clams, scallops, oysters, squid, octopus, and their relatives (fig. 16.1). Except for slugs, squid, and octopus, most molluscs have hard shells made of calcium carbonate, usually calcite but also sometimes aragonite (mother of pearl), which makes them easy to fossilize. Their fossil record is excel-lent, better than that of any other marine fossil group. More than 60,000 fossil species are known, and more paleontologists work on molluscs than on any other invertebrate animal group, most of them specialists in just one class, such as the snails or clams or ammonoids.

Molluscs are also important to human culture. Seashell collecting is not only a popular pastime now but has been for centuries, and in many cul-tures shells were used as money ("wampum" to Native Americans). People in many cultures eat a wide variety of "shellfish," especially clams, oysters, and scallops, as well as abalone, conch, squid (*calamari*), octopus, and land snails (*escargot*). Nearly all prehistoric peoples who lived near the coast were big mollusc eaters, as shown by their large garbage heaps known as shell middens. Over 2 million tons of squid and octopus are harvested each year, along with enormous volumes of clams, oysters, and scallops. In the United States alone, more than 220 million pounds of clams, oysters, and scallops are eaten each year. Finally, certain molluscs (especially oysters) produce a valuable gem, the pearl, from the aragonite that their mantle secretes to any irritant that gets inside the mantle cavity.

Figure 16.1 ▲

Molluscs come in many different shapes and forms, from the creeping chitons of the tide pools (*upper left*), to the two-shelled bivalves such as clams and oysters and their kin (*upper right*), the spiral-shelled snails or gastropods (*lower left*), and the highly intelligent, fast-moving, tentacled cephalopods such as this cuttlefish (*lower right*). (Courtesy of Wikimedia Commons)

In addition to being extremely diverse, molluscs have come to inhabit almost as many ecosystems as the arthropods. They live in every marine setting from the deepest ocean to the intertidal zone, and some are planktonic (float on the open sea). Many different freshwater clams and snails are known, and land snails are well adapted to being away from water.

Molluscs achieve this flexibility despite having a rather limited body plan. Nearly all molluscs are built around a softy fleshy body with a muscular foot that helps propel them, and with a shell on their back. Their body is covered by an organ called the *mantle*, which secretes their shell. Most molluscs have a simple digestive tract that runs from mouth to anus, paired gills for breathing in water, and simple versions of most other systems, including circulatory, excretory, reproductive, and other systems. Although most molluscs tend to be small, some are huge. The giant squid allegedly reaches about 50 feet (18 meters) in length, the giant clam is over a meter long, and the giant marine snail *Campanile* has a shell more than a foot long.

The simplest molluscs are the limpet-like *monoplacophorans*, which have a cap-shaped shell over their back, but they still have segmented muscles, gills, and other body parts just like their relatives, the segmented worms. Monoplacophorans were the most common molluscs in the Early Cambrian. It was thought that they had become extinct by the middle Paleozoic, but living examples of monoplacophorans were found in deep oceanic trenches in 1952.

From this basic body plan of the "hypothetical ancestral mollusk," the various classes of molluscs have modified their anatomy to live in different ways (fig. 16.2). Among the least modified are the *chitons* (class Polyplacophora or Amphineura), which are familiar denizens of tide pools around the world (see fig. 16.1, upper left). In the intertidal zone, they stick to rocks with their broad muscular foot, and they have modified the primitive limpet-like shell by subdividing it into eight plates, which enables them to bend and flex their body to fit around curves as they cling to rocks. Like limpets, they spend their lives slowly creeping along, scraping algae with their ribbon of tiny teeth made of iron oxide, or magnetite. In some islands in the tropical Pacific, the chitons have scraped so much rock away from the tide-pool zone that the islands are on pedestals.

Another minor group of molluscs are the tusk shells (fig. 16.2), or *scaphopods* (class Scaphopoda). Most of them have a long conical shell with a hole at the tip that resembles an elephant tusk, hence their name. These animals burrow in the shallow sand in the near-shore zone with their open tip exposed, allowing them to filter ocean water and trap food while releasing their waste products. In addition, a number of extinct molluscan classes, such as the rostroconchs, are rarely fossilized and are mostly known from the Cambrian, when molluscan evolution was undergoing its "experimental" stage.

Three classes of molluscs make up the bulk of both the living and fossil species and deserve detailed discussion. These are the Gastropoda, the Bivalvia, and the Cephalopoda.

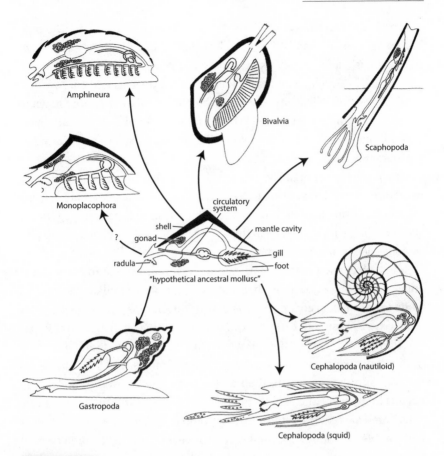

Amphineura

Bivalvia

Scaphopoda

Monoplacophora

circulatory
system

shell

gonad

?

radula

mantle cavity

gill

foot

"hypothetical ancestral mollusc"

Cephalopoda (nautiloid)

Gastropoda

Cephalopoda (squid)

Figure 16.2 ▲

Radiation of the molluscs from the "hypothetical ancestral mollusc." (Redrawn by Mary Persis Williams from several sources)

CLASS GASTROPODA (SNAILS, SLUGS, AND NUDIBRANCHS)

The Gastropoda are by far the most diverse of all molluscan classes, with over 100,000 species, making up about 80 percent of the living molluscs. The simplest molluscs are the limpets and their relatives, which have not changed that much from the monoplacophorans and other archaic molluscs. They are subdivided into many groups. Most of the marine snails are known as *prosobranchs* ("forward gills" in Greek), and they rotate their gills from the rear to over their head during larval development. The bubble shells, plus the snails that have lost their shells (sea hares, sea slugs, nudibranchs), and the planktonic pteropods are *opisthobranchs* ("backward gills" in Greek). The third group is the land snails and slugs, which have turned their gill cavity into an air-breathing organ; they are known as *pulmonates* ("lunged" in Greek). The opisthobranchs and pulmonates have only a limited fossil record, so I will focus on the abundant shell record of prosobranchs here.

Among the prosobranchs, the simplest are the limpets and abalones, which are adapted to sticking to hard surfaces and scraping off algae using the ribbon of tiny teeth in their mouth. But the simple, cap-shaped limpet shell does not allow for much in the way of evolutionary innovation, so gastropods soon evolved into other forms.

If a conical shell gets too large, it is much more stable to carry if it is coiled into a spiral, and nearly all of the other gastropods have evolved some variation on a coiled shell. The basic terminology of a coiled shell is quite easy to remember (fig. 16.3). Each turn of the shell around the axis is called a *whorl*. The number of whorls and the width and rate of turning of the shell are diagnostic of certain groups of snails, and they change as the snail grows larger in its shell and adds more material to the leading edge. The point of the top of the shell is the *apex*, and the shell can be considered *high-spired* if it is long and pointed, or *low-spired* if it is shorter and stubbier. If you see the shell broken open, the internal column formed by the junction of the inner part of each whorl is called the *columella*. The opening (*aperture*) of the shell typically has an *inner* and *outer lip*, and an area next to the inner lip called the *callus*, where the shell rested on the back of the snail. Many advanced gastropods have a notch at the bottom of the shell called the *siphonal notch*. Finally, coiled shells are asymmetric and can be either right-handed (*dextral*) or left-handed (*sinistral*). If you hold the shell with the spire pointed upward, the aperture will be on the right in a dextral shell, and on the left in a sinistral shell. Most snails are dextral, although some are sinistral, and a few snails change their coiling direction from one to the other.

However, there are many ways of making a coiled shell. During the early Paleozoic, there were many experiments in other types of shell coiling among the gastropods. Some, like the bellerophontids, coiled their shells in a flat spiral that stood symmetrically over the middle of their back.

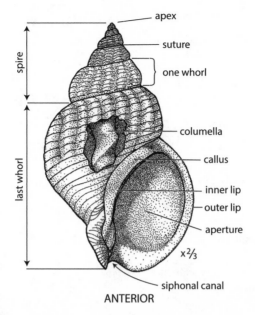

spire

apex

suture

one whorl

columella

callus

inner lip

outer lip

aperture

x²⁄₃

last whorl

siphonal canal

ANTERIOR

Basic anatomy of the gastropod shell. This is a dextral (right-handed) shell, which has its aperture to the right in this standard orientation. A sinistral (left-handed) shell would be the mirror image of this. (Illustration by Mary Persis Williams)

Bellerophon is a fairly widespread fossil in many Paleozoic localities. Another way to arrange the shell is to hold the spiral diagonally across the back and pointed forward and over the head, an arrangement called *hyperstrophic*. This early experiment in shell evolution is best seen in the common large Ordovician snail known as *Maclurites* (fig. 16.4).

These early experiments in shell evolution vanished by the late Paleozoic, and all other prosobranch gastropods have their shell positioned diagonally over their back with the apex pointed backward; this is the *orthostrophic* condition. In addition to the limpets and abalones and other primitive gastropods, many coiled gastropods still retain their primitive arrangement of gills, such as the living slit shells, top shells, turban shells, dog whelks, and the sundial shell, *Architectonica*.

Slightly more advanced snails are known as *mesogastropods*, a group containing about 30,000 species (fig. 16.4). They have reduced their paired gills down to a single unbranched gill; they also tend to have a much

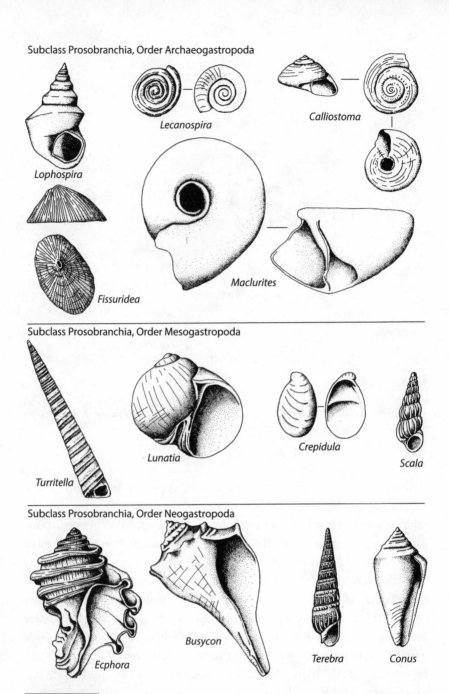

Subclass Prosobranchia, Order Archaeogastropoda

Lophospira

Lecanospira

Calliostoma

Fissuridea

Maclurites

Subclass Prosobranchia, Order Mesogastropoda

Turritella

Lunatia

Crepidula

Scala

Subclass Prosobranchia, Order Neogastropoda

Ecphora

Busycon

Terebra

Conus

`Figure 16.4` ▲

Some typical archaeogastropods (*top*), mesogastropods (*middle*), and neogastropods (*bottom*). (Redrawn by Mary Persis Williams from several sources)

narrower shell opening. These include the very high-spired shells such as the turritellids, plus periwinkles, moon snails, slipper shells, and many others, including a number of freshwater snail groups.

The most advanced snails are distinguished by having modified their mantle cavity into a long siphon that serves like a snorkel to allow them to breathe fresh seawater while burrowing beneath the surface. These snails are known as *neogastropods*, and their shells all have some kind of distinct siphonal notch, and often a long flange where the siphon is protruded out of the shell (fig. 16.4). The long siphon allows many of them to burrow into the sediment, with only their siphon exposed like a snorkel. They originated in the Early Cretaceous and soon became very abundant in the fossil record, with more than 16,000 species alive today. Most of the familiar marine snails such cone shells, conchs, cowries, whelks, muricids, olive shells, mud snails, volutes, turrids, auger shells, nutmeg shells, and hundreds more, are neogastropods.

CLASS BIVALVIA (= PELECYPODA, = LAMELLIBRANCHIA): CLAMS, OYSTERS, SCALLOPS, AND THEIR RELATIVES

Clams, scallops, and oysters are familiar to us all. They are the second most common group of living molluscs, with about 8,000 to 15,000 living species; they include a huge number of fossil species as well (at least 42,000 species). They are formally known as Bivalvia because the original molluscan cap-shaped shell of a limpet is modified into two shells, or valves, that hinge with each other. In many books, you may find them called by an obsolete name, "pelecypods," which means "hatchet foot" in Greek. This describes the way their muscular foot is modified into a long probing appendage that can wedge itself into the sand beneath them and help them dig.

Bivalves have a very peculiar body plan (fig. 16.5). From the simple limpet-like ancestor, they have not only enlarged and hinged their shells over their bodies, but they have even lost their heads. Instead, nearly all of the internal volume of the shell is filled with their gills. An even older obsolete name is the "Lamellibranchia" ("layered gills" in Greek), which refers to how their gills were used not only for breathing but also for filter feeding. As they open their shell and let the seawater flow over the gills, they trap not only oxygen but also tiny food particles, which are then captured by mucus

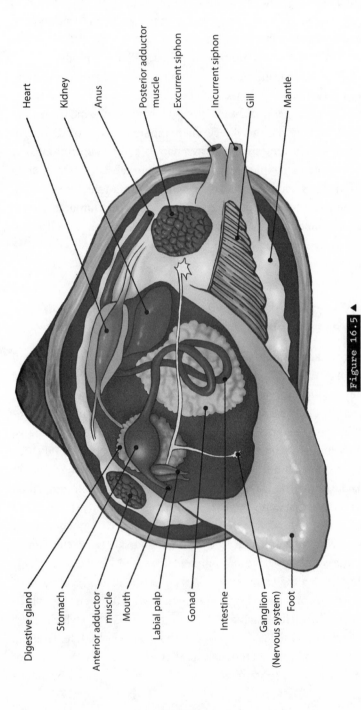

Heart

Kidney

Anus

Posterior adductor muscle

Excurrent siphon

Incurrent siphon

Gill

Mantle

Digestive gland

Stomach

Anterior adductor muscle

Mouth

Labial palp

Gonad

Intestine

Ganglion (Nervous system)

Foot

Figure 16.5 ▲

Basic anatomy of a clam. (Illustration by Mary Persis Williams.)

and flow toward the mouth at the base of the gills, and then through the digestive tract. The rest of the shell volume is taken up by reproductive organs, excretory organs, and muscles used to close the shell (adductor muscles). Finally, most bivalves have a large muscular foot that can dig down into the sand beneath them. Then the foot bulges its tip to anchor it as it pulls the shell beneath the sand. The clam rocks the shell back and forth as it knifes through the soft sediment, speeding up its sinking into the substrate. Many clams also pump water out of the shell cavity to liquefy the sediment into a soft slurry like quicksand, further accelerating the rate at which they disappear into the sediment and become less vulnerable to predators.

Most of these soft tissue features, however, are not preserved in the shell. Instead, paleontologists use the shape of the shell, details of the hinge with its teeth and sockets, and other aspects of the shell to identify the species. Bivalve shells are symmetrical *between* the valves (the right valve is the mirror of the left valve), in contrast to brachiopods, whose symmetry is *through* the valves (see fig. 13.2B). Although bivalves have many living positions, the standard anatomical orientation used by scientists to describe them is to point the hinge upward, with the shorter end of the shell away from your body. When you do this, the right valve will be in your right hand and the left valve will be in your left hand. In this position, the foot and the mouth are on the front edge (anterior) of the shell (pointed away from you), and the long mantle tube called the siphon is at the rear (posterior) of the shell (pointed toward you).

Other useful landmarks on the inside of a single shell can help us understand their function and aid in identification (fig. 16.6A–G). Bivalves have *ligaments* in the hinge area that spring the hinge open automatically when their adductor muscles, which close the shell, relax. This is the reason clams open automatically when they die, and it also explains why their fossilized shells are usually found separated from each other. The roughened areas where the muscles once pulled on the shell are known as *adductor muscle scars*. Most clams have two adductor muscles, an anterior and a posterior adductor, so there will be two scars inside each shell. Scallops are peculiar in that they swim by clapping their valves like castanets, which creates jets of water that enable them to swim erratically and escape predators. To do this, they have only one strong column of muscle, and only a single muscle scar on each shell. That muscular column is what you eat when you are served scallops in a restaurant.

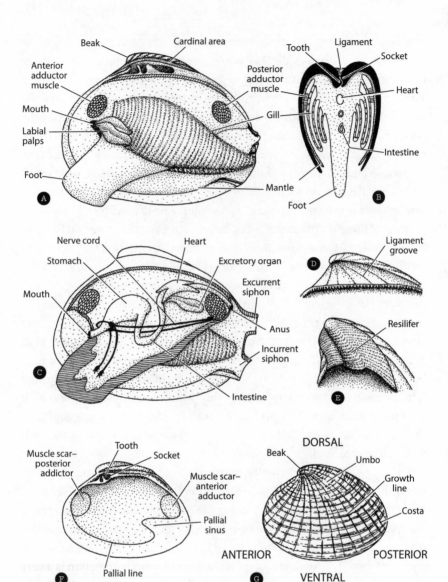

Figure 16.6 ▲

(A) Lateral view of a bivalve with the left valve and mantle fold removed; (B) cross-section of bivalve with shell shown in black; (C) lateral view of interior of right valve; (D) cardinal area showing ligament groove; (E) cardinal area with resilifer; (F) lateral view of interior right valve; and (G) lateral view of exterior of left valve. (Redrawn from several sources by Mary Persis Williams)

Another feature of the inside of the shell is a boundary line dividing the area where the mantle contacted the shell (nearer the hinge) with areas where the mantle did not contact the shell (fig. 16.6F); this is called the *pallial line*. Typically, on the posterior part of the pallial line is an irregular notch called the *pallial sinus*. This is where the siphon protruded through the mantle: the deeper the sinus, the larger the siphon.

Important anatomical features are also found on the outside of the shell (fig. 16.6G). Above the hinge is the bulge or hump at the top of the external shell, called the *umbo*. The *beak* is where the shell spirals to a point. The *growth lines* spread concentrically from the hinge area, and the *costae* or ribs radiate out from it like spokes in a wheel. Some shells, such as scallops, have large corrugations or folds on the shell, known as *plications*. As in brachiopods, the line of the opening of the shell is called the *commissure*. Most clams that burrow must maintain a relatively smooth shell to reduce drag, but oysters and scallops and others that only sit on the bottom and do not burrow may have an irregular outer surface, often with spines (such as the thorny oyster), which makes it harder for a predator to break them in order to feed on them.

I cannot discuss all of the thousands of species and hundreds of genera of bivalves here, but it is easy to tell something about the life habits of a bivalve by looking at the shell (fig. 16.7). Scallops (subfamily Pectinaceae), for example, have a distinctive shape (familiar as the Shell Oil Company logo). Scallops live on the seafloor with their valves open and have a dozen tiny blue eyes on the edge of their mantle. When they detect a threat, they jet-propel themselves above the seafloor to escape by flapping their valves together, producing a jerky irregular movement that confuses most predators. (Search for videos of scallops swimming online—it is amazing to watch.) Oysters (subfamily Ostraceae), on the other hand, sit on the seafloor and give up any attempt to burrow or swim. Their protection is a very thick shell that is hard to open or break. In addition, many oysters live in brackish water that fluctuates between normal marine salinity of 3.5 percent and fresh water, which is an environment that most of their predators can't stand. Mussels (subfamily Mytilaceae) live in rocky tide pools and attach to the rocks with thick, strong, fibers called *byssal threads*, which prevents predators from pulling them off the rocks.

Burrowing bivalves come in lots of different shapes and sizes. The most common type are the familiar venus clams (subfamily Veneraceae), some of

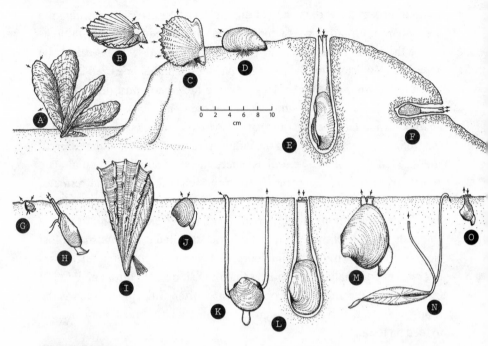

Figure 16.7 ▲

Ecology of bivalves. Epifaunal suspension feeders: (*A*) the cemented oyster *Crassostrea*; (*B*) the swimming scallop *Pecten*; (*C*) byssally attached pearl oyster *Pinctada*; and (*D*) the mussel *Mytilus*. Infaunal bivalves: (*E*) the angel wing *Pholas* and (*F*) the geoduck *Hiatella*, siphonate suspension feeders that bore into rocks; (*G*) the nut clam *Nucula*, a nonsiphonate labial palp deposit feeder; (*H*) the nut clam *Yoldia*, a siphonate labial palp deposit feeder; (*I*) the pen shell *Atrina* and (*J*) the clam *Astarte*, nonsiphonate suspension feeders; (*K*) the lucinid *Phacoides*, an infaunal mucus tube feeder; (*L*) the soft-shelled clam *Mya* and (*M*) the venus clam *Mercenaria*, siphonate suspension feeders burrowed into the sediment; (*N*) the tellin *Tellina*, a siphonate deposit feeder; and (*O*) the septibranch dipper clam *Cuspidaria*, a siphonate carnivore. (After *Geological Society of America Memoir* 125, by S. M. Stanley. Reproduced by permission of the publisher, the Geological Society of America, Boulder, Colorado © 1970 Geological Society of America)

them are called quahogs. Most of them are shallow burrowers living in the shallow subtidal seafloor, although some species live in the surf zone and can burrow down out of sight in a matter of seconds. They have relatively short siphons, and this is reflected in the relatively shallow pallial sinus on the shell. Other clams, such as the lucinids (subfamily Lucinaceae) and the tellins (subfamily Tellinaceae), are deeper burrowers, with very long siphons. Some bivalves are very deep burrowers. Their thick siphons are so large that they cannot close around them, leaving a permanent gape in the posterior of the shell. These include the soft-shelled clams (subfamily Myaceae), such as the mud-burrowing geoduck (pronounced "gooey-duck") *Panopea*. They may have a thick siphon as long and thick as your arm, with a small shell at the end. Other deep burrowers with long siphons are the razor clams (family Solenidae). A number of clams have adapted for rock boring (family Pholadidae), and their shell is reduced to a pair of cutting blades at the end of a long fleshy body that is protected by their burrow instead of by a shell.

Some extreme variations on the basic clam body shape lose their symmetry when they become oyster-like. For example, the giant clams (family Tridacnidae) have a huge thick shell that is corrugated but seldom closes completely (fig. 16.8A). This gap is covered with a thick area of mantle filled with symbiotic algae in their tissues, which help them grow to such giant sizes (just as algae allow corals to grow their large reefs of limestone).

A number of interesting extinct oysters are also common in fossil collections. A common Triassic fossil is the triangular clam known as *Trigonia*, whose name means "triangular" in Greek (fig. 16.8B). Trigoniids survived most of the Mesozoic but were thought to be extinct until a living descendant, *Neotrigonia*, was discovered in 1802 in the South Pacific, a true living fossil. During the Jurassic, one group of oysters grew into a spiral with one valve and a small lid on the other (fig. 16.8C). Known to paleontologists as *Gryphaea*, their common name is "Devil's toenails." In the Cretaceous, an even weirder group of oysters evolved with a corkscrew spiral shape that resembles a snail shell in one valve and a lid on the other (fig.16.8D–E). It is known as *Exogyra*, which translates from the Greek as "spiraling outward."

Another group of Cretaceous clams were the huge flat inoceramids (fig. 16.8F–G). Both valves were shaped like shallow plates, and the biggest ones were 5 feet (1.5 meters) across. They apparently lay flat on the seafloor of the shallow waters of the Cretaceous seaways. Many are found with complete fish and other fossils inside them, suggesting that they harbored a number of symbiotic organisms that sought shelter inside their huge shells.

But the most bizarre of all bivalves were the Cretaceous reef-building oysters known as rudistids (fig. 16.8H–I). One valve was shaped like a large cone, embedded point down into the seafloor. The other valve was shaped like a lid that covered the opening, and it would open whenever the clams needed to let in currents for food and oxygen. Rudistids formed huge reefs of densely packed shells in the tropical waters of the Cretaceous and even drove corals from the reef region.

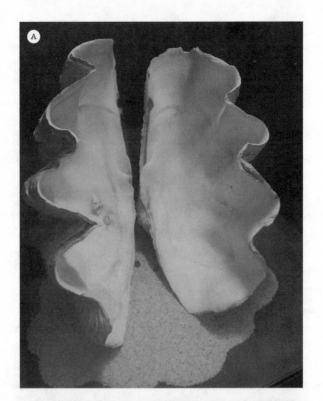

Figure 16.8 ▲

A variety of odd-shaped bivalves, especially oyster-like forms that abandon all attempts at bilateral symmetry because they do not burrow. (*A*) The living giant clam, *Tridacna*, which spends its time in coral reefs using algae in its mantle tissues to help it grow. (*B*) The triangular Triassic clam, *Trigonia*. (C) The coiled dish-shaped *Gryphaea*, also known as "Devil's toenails." (*D–E*) The spirally coiling Cretaceous oyster *Exogyra*. (*F*) The giant flat "dinner plate" clam, *Inoceramus*. (*G*) Reconstruction of inoceramids as they may have looked on the Cretaceous seafloor. (*H–I*) Two examples of the huge cone-shaped rudistid oysters, which built the major reefs during the Cretaceous. ([*A, C–I*] Photographs by the author; [*B*] courtesy of Wikimedia Commons)

1.0 cm

Figure 16.8 ▲

(*continued*)

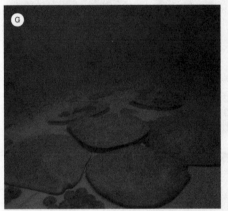

Figure 16.8 ▲
(continued)

CLASS CEPHALOPODA: SQUID, OCTOPUS, CUTTLEFISH, NAUTILOIDS, AND AMMONOIDS

Molluscs have evolved a wide variety of body plans, from the slow-moving snails to the headless clams that burrow beneath the surface and filter feed, to the chitons and tusk shells, and to the planktonic pteropods. But the champions of molluscs in terms of speed, intelligence, and complex behavior are the cephalopods. Cephalopoda means "head foot" in Greek, which refers to how the primitive foot of molluscs has been modified into a ring of tentacles around the head and mouth.

Speed and intelligence are the hallmarks of the group. Squids jet-propel themselves backward through the water faster than most animals can swim. The octopus is legendary for its intelligence, amazing quickness, and camouflage ability that enables its skin pattern to change. In fact, many cephalopods can rapidly change their skin patterns and even flash patterns in their skins faster than neon lights, which they use for communication. Most cephalopods also have an amazingly well-developed eyeball that is much like the vertebrate eye, but it evolved independently and converged on the vertebrate eyeball. It has a cornea, lens, and retina, but it is much better designed than our eyes. Octopus eyes don't have a blind spot in the retina for the exit of the optic nerve, nor is the sensory layer of the retina covered by blood vessels and other tissues that distort the light, as in vertebrate eyeballs. Vertebrates also have their photoreceptors pointed backwards in the retina, away from the light source.

Most cephalopods are large active predators or scavengers, floating above the seafloor in search of prey that they grab with their tentacles armed with suckers. Most of them expel water from the mantle cavity to give them a sort of jet propulsion, especially when they are trying to escape predators. Some, such as squid and octopus, even leave a cloud of ink behind them as a smokescreen to cover their escape. In addition to their water jets, squid and cuttlefish can swim slowly forward using the fins along the side of their bodies, and octopus mostly move with their long arms.

Cephalopods range in size from tiny squid only a few inches long to the giant squid that was about 50 feet (18 meters) long (fig. 16.9A). The maximum leg span of an octopus is more than 33 feet (10 meters). Some ammonite shells were more than 5.2 feet (1.7 meters) across, suggesting that the soft body was as large as the largest squid today (fig. 16.9B). Cephalopods

Giant cephalopods. (A) The giant squid, which lives in the deepest part of the ocean, and rarely washes ashore. They reach about 50 feet (18 meters) in length. (B) The gigantic ammonite *Parapuzosia*, which would have had an enormous head and long tentacles protruding from its opening. ([A] Courtesy of Wikimedia Commons; [B] photograph by the author)

Color Figure 1 ▲

These Cretaceous ammonites still have their rainbow-colored iridescence from "mother of pearl" aragonite preserved in their shells (see also fig. 2.1D). (Photograph by the author)

Color Figure 2 ▲

This ammonite shell has been completely replaced by pyrite (see also fig. 2.3B). (Courtesy of Wikimedia Commons)

Color Figure 3 ▲

The living inarticulate brachiopod *Lingula*, shown on the sediment surface after having been extracted from its burrow (see also fig. 13.5). (Courtesy of Wikimedia Commons)

Color Figure 4 ▲

Restoration of the bryozoan *Archimedes* in life (see also fig. 14.3D). (Illustration by Mary Persis Williams)

Color Figure 5 ▲

Occasionally, black shales preserve the soft parts of trilobites, here replaced by pyrite. This specimen shows the legs, gills, and antennae (see also fig. 15.5B). (Courtesy of Wikimedia Commons)

Diorama of the orthocone nautiloid as it would have appeared in the Ordovician (see also figure 16.12B). (Photograph by the author)

Diorama of the Cretaceous seafloor showing a large normally coiled ammonite, a corkscrew-spiraled heteromorph, the long straight *Baculites*, and the squid-like belemnites (see also fig. 16.14I). (Photograph by the author)

A typical graptolite colony of *Diplograptus* and *Dictyonema* as they might have looked floating on the Ordovician sea surface (see also fig. 18.3). (Illustration by Mary Persis Williams)

Head shield of *Pteraspis* (see also fig. 19.2C). (Courtesy of Wikimedia Commons)

Color Figure 10 ▲

Head and body shield of the giant arthrodire *Dunkleosteus* (see also fig. 19.3A). (Courtesy of Wikimedia Commons)

Color Figure 11 ▲

Reconstruction of the school-bus-sized Paleocene anaconda *Titanoboa* (see also fig. 19.11G). (Photograph by the author)

The giant brachiosaur sauropod *Giraffatitan*, on display in Museum für Naturkunde in Berlin (see also fig. 19.16D). (Courtesy of Wikimedia Commons)

▲

The primitive rhino-like group known as brontotheres (see also fig. 19.24B). (Photograph by the author)

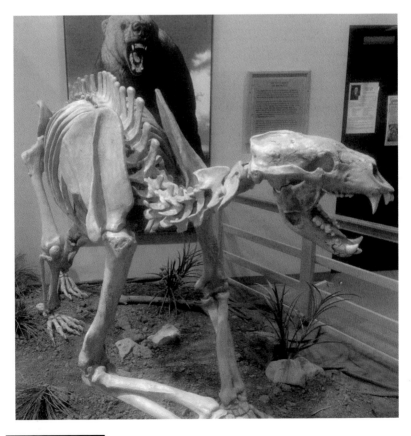

▲

The short-faced bear, one of the largest land predators that ever lived; it is bigger than any living bear (see also fig. 19.30G). (Photograph by the author)

live exclusively in marine waters; they were unable to colonize the freshwaters or move on land as snails and clams did. However, they sometimes live in huge numbers in the ocean, with some squid schools containing more than a thousand individuals. Many cephalopods, like the vampire squid, specialize in very deep, dark, cold parts of the ocean from 1.8 to 3 miles (3,000 to 5,000 meters) below the surface.

There are about 1,000 living species of squids, octopus, and cuttlefish; they have no external shell, although squids and cuttlefish have a rod of a hard substance (calcite or an organic material) holding their bodies rigid. Only one group of living cephalopods has an external shell, the famous chambered nautilus of the South Pacific (fig. 16.10A–C). They are practically

Figure 16.10 ▲

The chambered nautilus, the only living shelled cephalopod. (*A*) A living *Nautilus*, showing the tentacles, eye, and hood over the opening. (*B*) Diagram of the external features of the living *Nautilus*. (*C*) Cross-section of the anatomical features of the interior of the *Nautilus*. ([*A*] Courtesy of the Wikimedia Commons; [*B–C*] redrawn from several sources by Mary Persis Williams.)

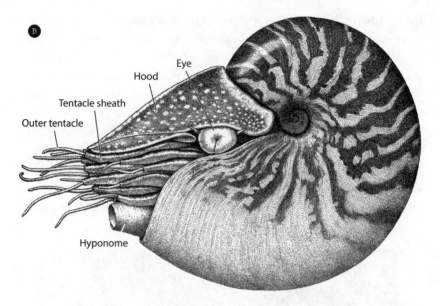

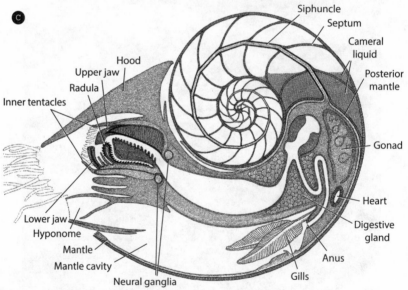

Figure 16.10 ▲
(*continued*)

the only living analogue for the 17,000 species of extinct shelled cephalo-pods in the fossil record. There are many different types of extinct nauti-loids as well. One extinct group that branched off from the nautiloids is the ammonoids, one of the most successful of all groups in the fossil record. They got their name because their coiled shells resembled the horns of the Egyptian ram-god Ammon. In the Middle Ages, some were thought to be petrified snakes (serpent stones), and the last part of the shell was some-times carved with a snake's head to enhance its serpentine appearance and market value. By 1700, the English naturalist Robert Hooke, the father of microscopy, saw the newly discovered chambered nautilus and correctly inferred that ammonites were related to nautilus, not snakes.

The nautilus shell is coiled in a flat spiral, like the snails that coil in a plane. However, unlike the shell in snails, the cephalopod shell interior is divided into chambers by a series of walls called *septa* (fig. 16.10C). The mollusc itself lives only in the last chamber. As its shell grows, its body moves forward until it can seal off another portion of its shell with a new septum. Once the septum seals the chamber, the nautilus uses a long fleshy stalk from its body, called the *siphuncle*, that connects all the chambers to slowly pull the water out of the new chamber using osmosis, leaving it filled with gases that help the shell float. The blood in the umbilicus is much salt-ier than the water in the sealed chamber, so the water slowly diffuses into the umbilicus until only gas remains. Contrary to popular myth, the nauti-lus cannot pump water in and out of the old chambers rapidly to change its buoyancy and rise or sink in the depths of the ocean.

The edge of the septum meets the outer shell at a line of intersection called the *suture*. These are usually only visible in fossils if the outer layer of shell has been eroded or polished away, exposing the outer edge of the septum (fig. 16.11). The suture pattern is the most diagnostic feature of most cephalopod shells, and it helps define the different species, genera, and even higher groups. In the primitive nautiloids, it is simple broad curve. As we shall see when we look at the extinct ammonoids, the suture became more complex and intricate, and this complexity makes it possible to recog-nize and identify the major groups of ammonoids.

The rest of the body of the nautilus resembles the unshelled squid and octopus in shape. The large head is covered with a hood, and big eyes domi-nate body. Nautilus has a parrot-like beak in the mouth, surrounded by a ring of tentacles, which it uses to catch prey (mostly crustaceans and carrion).

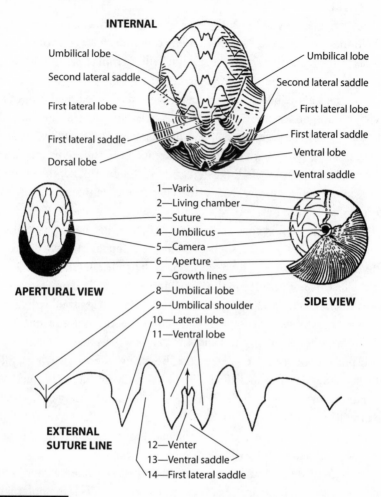

INTERNAL

Umbilical lobe
Second lateral saddle
First lateral lobe
First lateral saddle
Dorsal lobe

Umbilical lobe
Second lateral saddle
First lateral lobe
First lateral saddle
Ventral lobe
Ventral saddle

APERTURAL VIEW

1—Varix
2—Living chamber
3—Suture
4—Umbilicus
5—Camera
6—Aperture
7—Growth lines
8—Umbilical lobe
9—Umbilical shoulder
10—Lateral lobe
11—Ventral lobe

SIDE VIEW

**EXTERNAL
SUTURE LINE**

12—Venter
13—Ventral saddle
14—First lateral saddle

Figure 16.11 ▲

The intersection of the edge of the septum with the outer shell forms a line called the suture, which can be seen in fossils if the outermost shell layers are scraped or eroded away. The details of the patterns of saddles and lobes in the suture enable paleontologists to identify almost any ammonoid. (Redrawn from several sources by Mary Persis Williams)

Beneath the head is the mantle cavity, where the gills lie. At the entrance to the mantle cavity is a nozzle called the *hyponome*, which funnels and focuses the jet of water to propel them. As the body squeezes down on the mantle cavity, it forces the water out the hyponome and creates a jet of water. Like octopus, however, nautilus uses jet propulsion only for rapid motion, especially when escaping. Most of the time, they use their tentacles to creep along the seafloor in any direction, especially when they are hunting.

SUBCLASS NAUTILOIDEA

The Nautiloidea includes the living chambered nautilus and a variety of Paleozoic relatives with long straight shells. They first appeared in the Cambrian, and by the Ordovician they were abundant in nearly every marine assemblage. Some were huge (fig. 16.12A-B) and became the largest animals

Figure 16.12 ▲

Straight-shelled orthocone nautiloids were the largest predators of the early Paleozoic, with shells reaching 33 feet (11 meters) in length, with even more length in their tentacles. (A) The size of *Cameroceras* compared to a 6-foot-tall person. The tentacles and soft parts are reconstructed, so they could have been much longer. (B; color figure 6) Diorama of the orthocone nautiloid as it would have appeared in the Ordovician. ([A] Illustration by Mary Persis Williams; [B] photograph by the author)

in the ocean. The biggest was *Cameroceras*, with some shells reaching 33 feet (11 meters) in length; the squid-like creature that once lived in that shell must have been as large as the living giant squid. Slightly smaller is the common fossil *Endoceras*, which was 11 feet (3.5 meters) in length. Most, however, were much smaller, typically only a few inches to a foot long.

Straight-shelled (orthocone) nautiloids were superpredators of the Ordovician and Silurian, and they ate almost anything they could catch. The inside of their large heavy shells were counterweighted with calcite deposits that served as ballast, so they floated horizontally above the sea bottom. Otherwise, the gases in the chambers would have made the shell float point upward. They could not move fast to chase prey, so they must have ambushed prey that crept within reach of their tentacles. As the first large marine superpredators in the oceans, they may have influenced how trilobites became more specialized to avoid predation in the Ordovician (see chapter 15).

Different groups of nautiloids are distinguished by where they deposited the calcite ballast in their shell. The Endoceratoidea filled their shell with cone-shaped deposits of calcite called *endocones*, which were nested one inside the other like conical cups in the dispenser next to a watercooler. They are most common in Ordovician beds around the world, and they died out in the Silurian. Another Ordovician group was the Actinoceratoidea, which are recognized by calcite deposits filling the tube around the siphuncle (*endosiphuncular deposits*). Other groups had deposits of calcite within the chambers (*endocameral deposits*). Usually, the specimen must be sliced open, or eroded or broken naturally, to reveal these internal features. Straight-shelled nautiloids reached their climax in the Devonian, and specimens from the Silurian and Devonian of the Tindouf Basin of Morocco are so abundant that they are commercially mined in large quantities and sold all over the world.

Nautiloids then declined during the rest of the Paleozoic, and the large straight-shelled orthocone nautiloids vanished altogether. However, the primitive coiled forms persisted through the Mesozoic in small numbers, and they survived the Cretaceous extinction event much better than their relatives, the ammonites, which died out completely. The shell of a coiled nautiloid such as *Aturia* is occasionally found in Cenozoic beds all around the world, so they continued to survive despite losing the world's oceans to bigger predators such as fish.

SUBCLASS AMMONOIDEA

During the Devonian, one group of straight-shelled nautiloids (the bactrit-oids) is thought to have coiled up into a spiral in a flat plane and given rise to the next big radiation of cephalopods, the Ammonoidea. Ammonoids are distinct from nautiloids in that the siphuncle penetrates the septal wall on the outer rim of the shell (the *venter*) rather than in the middle of the septum, as in nautiloids (see fig. 16.11). From the Devonian through the Permian, there was a great evolutionary radiation of the most primitive ammonoids. They are known as *goniatites*, and they can be recognized by their distinctive zigzag suture pattern on specimens where the outer shell is removed (fig. 16.13A–B). They are excellent index fossils for the later Paleozoic, and in some localities (such as the Devonian beds of the Tindouf Basin of Morocco) the goniatite *Geisenoceras* occur in huge numbers and are commercially mined on a large scale and sold by shops and dealers all over the world.

During the great Permian extinction event, all but two lineages of goni-atitic ammonoids were wiped out. These survivors were the ancestors of a new diversification of ammonoids during the Triassic. These ammonoids had a distinctive *ceratitic* suture, which typically had a series of U-shaped curves in it (fig. 16.13A, 16.13C). At their peak, there were about 80 families and more than 500 genera of ceratitic ammonoids. The ceratitic ammonoids went through another extinction crisis at the Triassic-Jurassic boundary.

The survivors of this extinction underwent a final huge evolutionary radiation in the Jurassic and the Cretaceous. By the Jurassic, there were 90 different families, and even in the Cretaceous there were still 85 families, although they declined to only 11 families near the end of the Cretaceous. These ammonoids had a complex, intricate florid suture pattern (fig. 16.13A, 16.13D), known as an *ammonitic suture*. The details of their complex sutures allow the paleontologist to identify any good specimen to species.

During this great Mesozoic diversification, ammonites evolved into many shapes. Some were streamlined with sharp edges such as the *oxyconic* shells of *Placenticeras* and related genera (fig. 16.13D), and they must have been relatively fast swimmers (although none were as fast as squids or fish). Others had short, fat *cadicone* shells, and they must have moved slowly along the bottom to catch their prey. Some had coils that resembled a coiled snake (*serpenticone*), as seen in the genera *Dactylioceras* and *Stephanoceras* (fig. 16.13E).

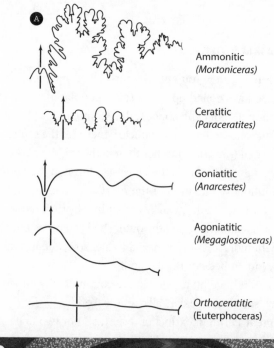

Ammonitic
(*Mortoniceras*)

Ceratitic
(*Paraceratites*)

Goniatitic
(*Anarcestes*)

Agoniatitic
(*Megaglossoceras*)

Orthoceratitic
(*Euterphoceras*)

Figure 16.13 ▲

The pattern of the sutures on the cephalopod shell is crucial to identifying them and determining the age of the beds from which they come. (*A*) Patterns of sutures on the shells through time, with increasing complexity from the simple curves of nautiloids to the convoluted sutures of ammonites. (*B*) Typical zigzag goniatite suture on a large *Geisonoc-eras* from the Devonian of Morocco. (*C*) The U-shaped ceratitic suture of the Triassic ammonoids called ceratites; this is the genus *Ceratites*. (*D*) The complex suture of ammo-nitic ammonoids of the Jurassic and Cretaceous seen in this *Placenticeras*. (*E*) The serpen-ticone ammonite *Dactlyioceras*. ([A] Redrawn from several sources by Mary Persis Williams; [B, D, E] photographs by the author; [C] courtesy of Wikimedia Commons)

Figure 16.13 ▲
(continued)

The most unusual ammonites, however, began to uncoil the simple flat spiral into a number of weird and bizarre forms. They are known as *heteromorph* ("different shape" in Greek) ammonites (fig. 16.14A). Some were just slightly uncoiled, like *Scaphites* (fig. 16.14B), or partially uncoiled like *Audoliceras* (fig. 16.14C). Others uncoiled into huge hairpin shapes, like *Hamites* and *Diplomoceras* (fig. 16.14D), weird question-mark shapes

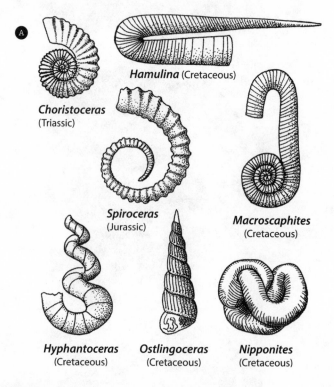

Ⓐ

Hamulina (Cretaceous)

Choristoceras (Triassic)

Spiroceras (Jurassic)

Macroscaphites (Cretaceous)

Hyphantoceras (Cretaceous)

Ostlingoceras (Cretaceous)

Nipponites (Cretaceous)

`Figure 16.14` ▲

Heteromorph ammonites started with simple flat spirals but became uncoiled in many weird and different ways: (*A*) some of the common heteromorph patterns; (*B*) slightly uncoiled *Scaphites*; (*C*) partially uncoiled *Audoliceras*; (*D*) hairpin-shaped *Diplomoceras*; (*E*) question-mark shaped *Didymoceras*; (*F*) straight-shelled *Baculites*, with some other heteromorphs; (*G*) coiled up an axis like a high-spired snail, such as *Turrilites*; and (*H*) *Nipponites*, with its shell twisted into a knot. (*I*; color figure 7) Diorama of the Cretaceous seafloor showing a large normally coiled ammonite, a corkscrew-spiraled heteromorph, the long straight *Baculites*, and the squid-like belemnites. ([*A*] Redrawn from several sources by Mary Persis Williams; [*B–D, G–H*] courtesy of Wikimedia Commons; [*E–F, I*] photographs by the author)

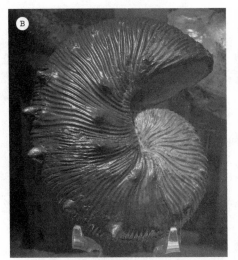

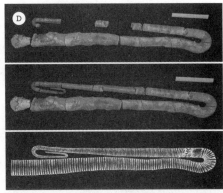

Figure 16.14 ▲
(*continued*)

Figure 16.14 ▲
(continued)

like *Macroscaphites* and *Didymoceras* (fig. 16.14E), uncoiled flat spirals (*Spiroceras*), and even stranger shapes. The oddest of all were those that coiled up a spiral axis like a snail such as *Turrilites* (fig. 16.14G), or coiled into a tight knot like *Nipponites* (fig. 16.14H). The heads of these ammonites were often in almost inaccessible positions, and they could not use jet propulsion because they would start to spin crazily. So they must have floated passively in the ocean and trapped nearby food with their long tentacles or crept slowly along the seafloor using their tentacles. Perhaps the most common and rapidly evolving heteromorphs were the baculitids (fig. 16.14F), which had secondarily straightened out their originally coiled shell into a nearly straight cone (with just a tiny coil sometimes remaining at the tip). Unlike the straight-shelled nautiloids with the counterweight ballast in their shells to hold them horizontal, baculitids had no ballast deposits inside them, so they floated with their shell point upward and their heads dangling below (fig. 16.14I). Clearly, they could not swim fast in this position, so they must have either crept along the seafloor using their tentacles or (as recent research suggests) floated in the open waters like plankton, trapping nearby prey (possibly even smaller plankton) with their tentacles.

After surviving the Permian extinction, the Triassic-Jurassic extinction, and several smaller crises, all of this great diversity of ammonites that had long dominated the Mesozoic seas (and are the best fossils for telling time) vanished with the great Cretaceous extinction. Only the coiled nautiloids and the cephalopods without external shells survived.

SUBCLASS COLEOIDEA

The coleoids include most of the cephalopods without external shells: squids, octopus, cuttlefish, argonauts, and their relatives. Without a hard external shell, they do not fossilize very well, and only a few complete specimens with soft tissues are known. However, one type of coleoid does fossilize well. Known as *belemnites*, they left a conical internal shell called a guard or rostrum, which reinforced their squid-like body and counterbalanced the weight of the head and tentacles. (In contrast, living squid have a thin, lightweight, flexible rod that serves to stiffen the back part of the body.)

Figure 16.15 ▲

Belemnites are the internal shells of an extinct group of squids. They produced abundant large calcite shells in the Jurassic and Cretaceous that looked like large bullets. (Courtesy of Wikimedia Commons)

Belemnite fossils look like large stone bullets (fig. 16.15), and they are particularly common in the Jurassic and Cretaceous, where some beds have concentrations of hundreds of belemnite fossils. A few belemnites have been found in extraordinary localities that preserve the soft tissue, so we know that these fossils came from a squid-like creature that is now extinct.

PHYLUM ECHINODERMATA

Echinodermata is the third largest phylum in the oceans after molluscs and arthropods. This phylum includes the sea stars, sea urchins, brittle stars, sea cucumbers, and crinoids or "sea lilies," and their extinct relatives (fig. 17.1A–E). Unlike the other two groups, however, echinoderms live only in normal marine waters and have not adapted to freshwater or life on land or in the air. They have no system for respiration or osmotic regulation, so they cannot survive without seawater surrounding them. Nevertheless, they have long been one of the most diverse and successful groups in the ocean. In the right circumstances, they can be incredibly abundant. Sea urchins are usually common in tide pools, along with sea stars. During the late Paleozoic, the seafloor was covered by huge meadows of crinoids and another extinct group called blastoids, representing many trillions of individual animals. The deepest part of the abyssal plain of the oceans— deeper than 3,300 feet (1,000 meters)—is dominated by echinoderms as well, especially sea cucumbers and brittle stars, where they make up about 90 percent of the marine life.

`Figure 17.1` ▶

(A) Sea stars may look pretty and ornamental, but they are very active and voracious pred-ators of all sorts of marine life, especially shelled molluscs. (B) Brittle stars are not predators, but they use their delicate arms to filter feed. (C) Sea cucumbers creep along the ocean floor feeding on the organic material in the bottom sediment. (D) Sea urchins creep along the bottom using their tube feet, defending themselves with hundreds of spines. (E) Crinoids or "sea lilies" have delicate feathery arms to filter-feed plankton, but they are attached to the bottom rather than freely roaming. (Courtesy of Wikimedia Commons)

Even though they are relatively advanced among the invertebrate phyla, echinoderms are truly strange and alien in their body form and construction. They start life as embryos and larvae with bilateral symmetry like most other invertebrates, but in their adult stages they give up the head and tail and become radially symmetrical (symmetrical around a central axis) like a sea jelly or a sea anemone. A few echinoderms, such as heart urchins and sand dollars, even superimpose some bilateral symmetry on top of this radial pattern. Not only are their larvae bilateral in symmetry, but the earliest echinoderms from the Cambrian were bilaterally symmetrical as adults, so the radial symmetry is a later development.

A second characteristic unique to echinoderms is that they are the only group of animals that employ hydraulics, like the hydraulic systems in large machinery, or the brakes on a truck or car. Hydraulic systems are built around an enclosed tube of fluid, which is used to transmit pressure from one area to another steadily over a long time. Likewise, the echinoderm body has a series of fluid-filled canals that are used to move them across the seafloor using *tube feet*, tiny fluid-filled extensions of the canal system with tiny suckers at the end. The hydraulic system means they move very slowly by our standards, but it also confers advantages. For example, a sea star feeds by wrapping its arms around a clam or mussel and then slowly pulling its shell open. The prey tries to hold its shell closed with its adductor muscles, but these muscles eventually become fatigued. The hydraulic system never tires, however, and can pull steadily and unrelentingly until the prey opens. At that time, the sea star turns its stomach inside out, inserts it inside the shell of the prey, and digests its victim in its own shell. The tube feet and hydraulics also give a sea star incredible grabbing power on rocks; it is almost impossible to pull them off or dislodge them.

With their loss of bilateral symmetry and their hydraulic bodies, echinoderms are truly weird, having lost many other systems we think of as normal in higher animals. Echinoderms have a nervous system, but no eyes, hearing, taste, or other specialized senses; they sense things by touch. The hydraulic system serves multiple functions, so echinoderms have no circulatory, respiratory, or excretory systems. This limits the echinoderms in important ways. Without an excretory system to regulate their water and salt balance, they must live in normal ocean water where salinity doesn't change much; there are no freshwater echinoderms. In addition, they don't vary in body size as much as molluscs or arthropods do.

The smallest echinoderms are at least a centimeter across, so there are no tiny planktonic forms (although echinoderm larvae are planktonic). A few sea stars are about 3 feet (1 meter) in diameter, and sea cucumbers may be 6.5 feet (2 meters) long, but none are larger. Some extinct crinoids had arms that flared out more than 3 feet (1 meter) and stems that were tens of feet in length, but they are nowhere near as bulky or massive as most giant marine animals.

Despite their weird construction, there is something even more surprising about echinoderms. Of all the marine invertebrate groups, they are the closest relatives of backboned animals or vertebrates, which include fish and humans. This is shown by many lines of evidence. For example, echinoderms and primitive marine relatives of vertebrates have the same kind of larva, known as a *tornaria* larva. When this larva first develops from a fertilized egg into a ball of cells with an opening at one end, that opening becomes the mouth of most invertebrates. But in echinoderms and vertebrates, that opening becomes the anus, and another opening develops to become the mouth (fig. 17.2). The cells of most invertebrates are determinate; that is, early in their development cells are specialized and cannot change function. However, the cells in an embryo of an echinoderm or vertebrate are indeterminate, and they can be split and develop into a different organ system or even become a completely new animal. The coelomic cavity of most invertebrates comes from the mesodermal wall, but in echinoderms and vertebrates it comes from the endoderm. Finally, analyses of the molecular sequences of echinoderms and other animals have shown over and over again that they are our closest relatives.

CLASSES OF ECHINODERMATA

The name *Echinodermata* means "spiny skinned" in Greek, and this name describes many of the groups. Most of the groups (especially sea urchins and sea stars) have spines that protrude from their shell or skin for protection. Echinoderms are built of a shell (*test*) composed of plates (*ossicles*) of single crystals of the mineral calcite. If you break an echinoderm ossicle, it will break along the cleavage planes of the calcite crystal. In most cases, they have a mouth at one end of the body, often surrounded by arms or tube feet, and the anus at the other end. There are five living classes of echinoderms.

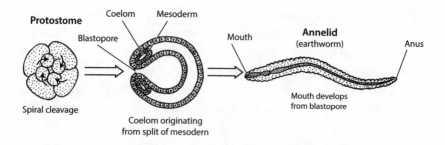

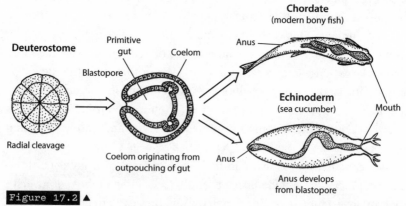

Figure 17.2 ▲

Comparison of the development of most invertebrate groups, the protostomes, which include worms and molluscs and arthropods (*top*), with the echinoderms and vertebrates (*bottom*), a group known as the deuterostomes. (Redrawn from several sources by Mary Persis Williams)

CLASS ASTEROIDEA

Sea stars are the most familiar of echinoderms to most people (see fig. 17.1A). They are often called starfish, but they are not fish in any sense, so biologists now call them sea stars. As described previously, sea stars are usually predators on other shelled marine animals, and they feed by inserting their stomach inside the victim's shell and digesting it. They have five arms (or arms in multiples of five) radiating around their mouth, which is on the bottom, and thousands of tube feet in their arms that give them a powerful grip on the rocks or on their prey. Their ossicles are small and embedded in their soft tissues, so when they die, their bodies tend to fall apart. Even though there are about 1,500 living species in about 430 genera today, they are rare in the fossil record.

CLASS OPHIUROIDEA

Brittle stars look much like sea stars, except their five arms are thin and delicate (hence the "brittle" in their name), and their body is a simple disk in the middle (see fig. 17.1B). The slender arms are not used for gripping; instead, they allow the brittle star to glide rapidly across the seafloor. Once they are in a good feeding place, they wave their arms in the current to trap tiny food particles. Even though there are about 2,000 living species in 325 genera, they are so delicate that they are fossilized rarely and only in extraordinary circumstances.

CLASS HOLOTHUROIDEA

Sea cucumbers are different from almost all other living echinoderms in that they have a long fleshy body with a mouth at one end and the anus at the other, and they have only tiny ossicles embedded in their body (see fig. 17.1C). However, tube feet run along the length of the body and help them move along the seafloor. Tube feet around the mouth help them shovel the organic-rich seafloor sediment into their mouth, where they digest the food material and pass the remains out their anus. They are found in many parts of the ocean and are by far the most abundant animals on the deep seafloor, where there is abundant organic detritus and few predators.

If a predator does disturb them, they have a truly bizarre defense mechanism. They extrude their internal organs (intestines, respiratory apparatus, and other systems) out their anus at the predator. When seawater hits these organs, they become a set of sticky threads that cover the intruder in nasty goo. After this happens, the sea cucumber can regenerate an entirely new set of internal organs.

Because they are made entirely of soft tissue, sea cucumbers rarely fossilize. However, they are very successful today, with 1,150 species in about 200 genera.

CLASS ECHINOIDEA

Sea urchins, heart urchins, and sand dollars have perhaps the best fossil record of any echinoderm class because they have a hard external shell, and many kinds already live in the seafloor sediment where they can become fossilized (see fig. 17.1D).

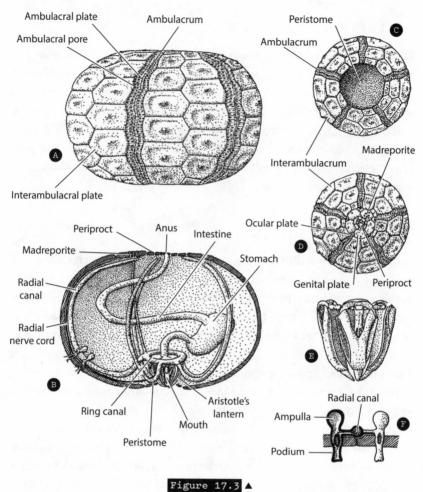

Figure 17.3 ▲
Anatomy of the sea urchin. (Redrawn from several sources by Mary Persis Williams)

Of all the echinoids, sea urchins have the most primitive body form (fig. 17.3). They have five rows of defensive spines in the *interambulacral areas* of the shell, alternating with five areas with rows of tube feet in the perforated *ambulacral areas* of the shell. Their long tube feet are used to creep along the surface and to move detritus off their shells. The mouth opening is at the bottom, and they have a set of five calcareous plates called *Aristotle's lantern* that serve as jaws to scrape algae off rocks or to feed on strands of kelp. The digestive tract then loops around the inside of the shell and out the top where the anus is located. The rest of the shell is filled with gonads (urchin roe, a delicacy in some places) and other internal organs, including their water vascular system. Sea urchins live in rocky tide pools and cruise around during high tide scraping algae. During low tide, they hide in small pockets in the rock that they have carved with their jaws.

By the Jurassic, the radially symmetrical sea urchins gave rise to other branches of echinoids that specialized in burrowing in the sediment rather than hiding in rocky tide pools. These urchins must burrow in a definite direction, so they have secondarily developed bilateral symmetry over-printing their radial symmetry. They have a front and back end, and right and left side, and are known as *irregular echinoids*. In most cases, this means that the mouth moves from the center of the bottom of the test to the front of the test, and the anus from the top of the test to the rear. With the pro-tection of a burrow, large defensive spines are no longer needed, so their spines have been reduced until they resemble a fuzzy coating. These spines are used to help them dig through the sediment. They no longer need jaws because they ingest marine sediment and digest the food material from it before passing it through their gut and out their anus.

Irregular echinoids have many different shapes. The first to evolve in the Jurassic were the heart urchins and sea biscuits, which have a tall, rounded shell (fig. 17.4A–E). These burrow into the sediment and use long tube feet to create a small respiratory canal to the surface of the sediment to bring them fresh seawater. Heart urchins were especially common in the Cretaceous where they lived in a variety of sedimentary settings. The most extremely modified are the sand dollars, which have developed a flattened body that helps them burrow just beneath the surface of the sed-iment (fig. 17.4F). When they are feeding, however, their tests stick out of the seafloor like a set of shingles, trapping currents on their underside and allowing the tube feet to convey the food particles to their mouth. The first sand dollars appeared in the Eocene from highly flattened Paleocene sea biscuits, and they soon spread to shallow marine seafloors worldwide. They are abundantly fossilized in many parts of the world and lived in a shallow marine setting with strong currents.

CLASS CRINOIDEA

Sea lilies are one of the most important groups of echinoderms in the Paleozoic, but they nearly vanished during the Permian extinction. Only a few groups have survived to the present day (see fig. 17.1E). Their bodies

Figure 17.4 ▲

Fossil echinoids: (A) nearly complete fossil of the long-spined regular urchin *Acrosalenia*; (B) two specimens of the regular urchin *Phymosoma*, one with spines still attached, the other showing the spineless shell; (C) shells of the sea biscuit *Tripneustes*; (D) Cretaceous heart urchin *Micraster*; (E) several Cretaceous heart urchins; and (F) dense cluster of fossil sand dollars, genus *Scutella*. (Courtesy of Wikimedia Commons)

have a head with arms on a long stalk rooted into the seafloor sediment, and they vaguely resemble a flower or a tree (hence the name "sea lilies"). However plant-like they may look, they are most definitely not plants, but animals related to sea stars and other echinoderms. (Search online for videos of "crawling crinoids" to see just how animal-like they really are.)

Only 25 living genera of stalked crinoids still survive, but there are about 6,000 species in 850 genera in the fossil record. These few surviving stalked crinoids are a tiny remnant of their former diversity, but they are all we have to study and understand the huge diversity of extinct crinoids. Most of them live in highly protected or sheltered habitats where there are few predators, and it is likely that the evolution of new predators in the Triassic explains why stalked crinoids never recovered the diversity or abundance they had in the Paleozoic. Enormous volumes of crinoidal limestone made entirely of broken crinoid ossicles demonstrate that during the Mississippian they lived by the trillions on the shallow seafloor across the entire continent.

Most of the living crinoids are a specialized form that has lost its stalk. They can crawl around on the seafloor or even swim short distances. They are called "feather stars" because their arms are feathery in appearance, and they vaguely resemble brittle stars and sea stars. About 700 species of these free-living crinoids swim in the oceans today.

The basic body (fig. 17.5A–B, fig. 17.6A) of a stalked crinoid consists of a "head" or *calyx* (filled with all the internal organs), which is topped by the mouth and surrounded by the arms around the mouth. The arms are covered with smaller branches called pinnules, and they form a large filter-feeding fan that traps food particles passing through them in the ocean currents. When the crinoid is feeding, the arms curl backward on themselves into a concave fan, like an umbrella, with the concave side facing the current to trap as much food as possible passing through their arms. Trapped food is then passed by sticky mucus down the arms and into the mouth. A short digestive tract processes the food and passes it to the anus near the mouth.

The calyx has a distinctive arrangement of plates (fig. 17.6B), which crinoid specialists use to identify the species and higher groups of crinoids. To identify species of these crinoids, you must learn the terminology for the plates in the calyx. At the end of the calyx opposite the mouth is the stalk of the crinoid, which is made of a set of rings of calcite (*columnals*) that

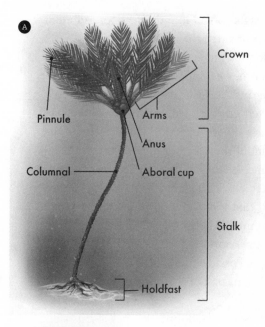

Figure 17.5 ▲

(A) Basic anatomy of a stalked crinoid. (B) A living stalked crinoid curls its arms back into the current like an umbrella to trap food-bearing water as it passes through the arms. ([A] Illustration by Mary Persis Williams; [B] courtesy of Wikimedia Commons)

resemble a stack of circular candies (fig. 17.6C). The stalk is held together by soft tissue in life, but when the crinoid died, the stalks usually fell apart. Huge volumes of limestone are made almost entirely of millions of broken crinoid columnals, the most easily recognized part of any crinoid. At the bottom of the stalk, most crinoids had a root-like set of structures called a *holdfast*, which anchored the crinoid in the sediment (fig. 17.6D). In fact, all of these parts disarticulated very easily when a crinoid died, so most fossils consist only of a calyx or columnals or a holdfast. Only in a few extraordinary localities is it common to find complete crinoids. Most crinoid specialists need only a decent calyx to identify the species because the stem and holdfast are seldom attached.

Crinoids were enormously successful during the Paleozoic, making up most of the marine biomass in many shallow marine settings, especially during the Mississippian, which has been nicknamed the Age of Crinoids.

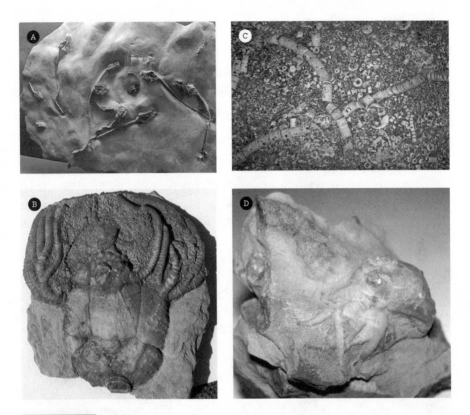

Figure 17.6 ▲

(*A*) Complete articulated crinoids like this show the complete anatomy of the animal, but these are rare. More commonly preserved are the individual parts of the animal. (*B*) Calyx, or head, of the crinoid *Platycrinites*. This one still has some of the arms, but most have lost the arms. (*C*) A crinoidal limestone, with both complete stems of crinoids and individual crinoid columnals, which resemble circular candies. (*D*) The rooting structure of the crinoid, called a holdfast. ([*A*] Photograph by the author; [*B*–*D*] courtesy of Wikimedia Commons)

However, as the seas retreated and the planet became colder in the Pennsylvanian and Permian, crinoids began to diminish as well, and they were severely depleted during the great Permian extinction. Only two or three lineages survived into the Triassic, and they never regained their former dominance, probably due to new predators in the seas that could eat defenseless crinoids. Today crinoids survive by being mobile (like the stalkless crinoids) or by living in hidden, protected areas with few predators (like the few remaining stalked crinoids).

OTHER EXTINCT CLASSES

There are as many as 20 to 25 other extinct classes of echinoderms, most of which are very rare and seldom encountered by fossil collectors. They are usually only collected and studied by specialists. However, one extinct class is very common (especially in the Mississippian), and it deserves mention here.

Class Blastoidea

Blastoids looked much like crinoids in that they had a "head" (called a *theca*) with arms at the end of a long stalk similar to Paleozoic crinoids (fig. 17.7A). However, the theca was constructed very differently than

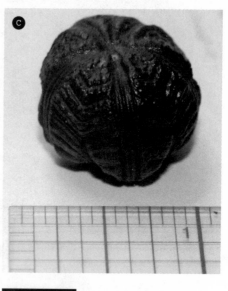

Figure 17.7 ▲

Blastoids. (*A*) Nearly complete head of the common Mississippian index fossil *Pentremites*, showing the arms still in place in the head of the blastoid. (*B*) Most of the time, blastoids are known only from their heads, such as this distinctive flower-bud shaped *Pentremites*, with its five-fold symmetry (hence the Greek word *penta*, meaning "five," in the name). (*C*) The globular blastoid *Orbitremites*. (Courtesy of Wikimedia Commons)

the calyx in crinoids. In most cases, the five-fold symmetry of the theca resulted in five ambulacral areas where the tiny thin tube feet and arms (usually lost in the fossils) once protruded. Inside the theca behind the ambulacral areas was a highly folded pleated structure known as the *hydrospire*, which apparently served for respiration. Some of the blastoids, like the ubiquitous Mississippian index fossil *Pentremites*, looked a bit like flower buds. Others were spherical in shape like *Orbitremites* (fig. 17.7C). Blastoids first appeared in the Ordovician, but their greatest diversity was in the Mississippian when they mingled with the immense shoals and meadows of crinoids. They declined again in the later Paleozoic and vanished completely in the great Permian extinction, along with trilobites, tabulate and rugose corals, and many groups of bryozoans, brachiopods, and crinoids that had dominated the Paleozoic.

PHYLUM HEMICHORDATA

The branch of animal life that includes backboned animals (vertebrates such as fish, amphibians, reptiles, birds, and mammals) has many close relatives. These relatives have some features of vertebrates, but they do not have key adaptations such as bone tissue or a spinal cord made of bony vertebrae. Nevertheless, they do have other key adaptations, such as the presence of a distinct throat region (*pharynx*), and most of them have a flexible rod of cartilage along the backbone, known as the *notochord*. Nearly all vertebrate embryos (including you when you were an embryo or fetus) have a notochord that is later replaced by the bony spinal column. These nonvertebrate animals with a notochord are part of the phylum Chordata to which all vertebrates also belong. Some of these chordate relatives include the tiny jawless fish-like group known as lancelets, and the tiny filter-feeding creatures known as tunicates (which have a tadpole-like larva with a notochord). None of these nonvertebrate chordates have much of a fossil record, so I will not consider them further here. However, they are important to understanding the transition from invertebrates to vertebrates.

The next closest relative to phylum Chordata is another group of soft-bodied marine creatures known as phylum Hemichordata, whose name means "half-chordates." The living hemichordates include a funny-shaped worm-like creature known as an acorn worm (which has a pharynx and other chordate features, unlike any living worm) and filter-feeding creatures known as pterobranchs (fig. 18.1A–B). Normally, these would only be the subject of an invertebrate zoology textbook. However, they turned out to be the crucial link in solving a persistent mystery in paleontology.

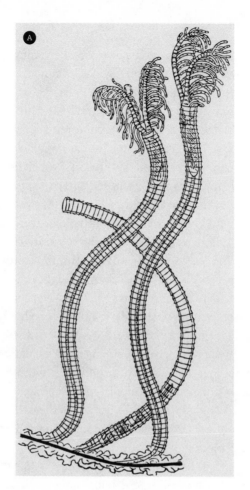

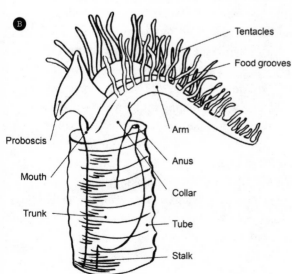

Tentacles

Food grooves

Proboscis

Mouth

Trunk

Arm

Anus

Collar

Tube

Stalk

Figure 18.1 ▲

The living pterobranchs are the closest relatives of graptolites. (*A*) Image of the pterobranch *Rhabdopleura*, showing the long tubular stolons and their distinctive wall structure. (*B*) The anatomy of the *Rhabdopleura*. (Redrawn from several sources by Mary Persis Williams)

Since the late 1700s, when geology was first becoming a science, scholars had puzzled over the strange-looking fossils found flattened on the surfaces of black deepwater shales. They were preserved as two-dimensional carbonaceous films on the bedding surface, so it was hard to visualize them in three dimensions. Only a film of graphite remained, without internal structures. It almost looked as if someone had marked the rocks with a graphite pencil, so they were called "graptolites," which means "written on stone" in Greek. Many ideas were proposed about the relationship of these animals to others, but there was little evidence to resolve the mystery.

Despite having no idea of what kind of biological creature they were, geologists were able to use graptolites for biostratigraphy. Graptolites evolved rapidly, and they were abundant both in deepwater shales that contained few other fossils and in shallow marine rocks. Soon they became excellent index fossils for the Ordovician, Silurian, and much of the Devonian. Graptolites were so useful for biostratigraphy that they helped solve the 50-year argument over the boundary between the Cambrian and the Silurian in the British Isles. Looking at the evolution of graptolites led to the proposal that there was a time period between the two disputed periods, which was named the Ordovician. More important, graptolites apparently floated around the world's oceans in the early Paleozoic, so it was possible to date any Ordovician, Silurian, or Devonian rock anywhere in the world using them.

The mystery of their form was not solved until 1948, when paleontologists found uncrushed three-dimensional graptolite fossils that were preserved in limestones and cherts. These fossils had not been flattened and altered, and by carefully slicing these rocks into thousands of tiny slices (and later by etching them in acid), their three-dimensional structure and anatomical details were finally revealed (fig. 18.2). Their detailed anatomy with rows of tiny cups (*thecae*) on a long branch (*stipe*) exactly matched the structure of the living pterobranchs still floating in the world's ocean. For this reason, graptolites are now considered to be a group of extinct hemichordates. If the extinct creatures were anything like their living relatives, then each little cup-like theca housed a tiny filter-feeding creature with a fan of tentacles near the mouth for trapping plankton, which then go through their pharynx and digestive tract. Modern pterobranch colonies (called *rhabdosomes*) apparently supported hundreds of individual animals on each stipe, connected to each other down the center by a thread of tissue

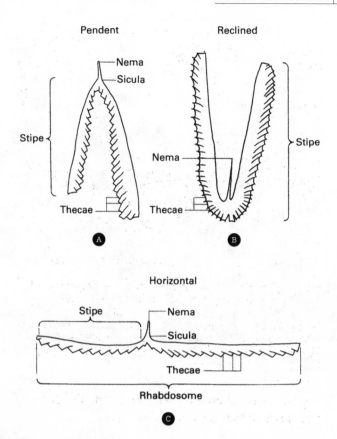

Figure 18.2 ▲

Detailed anatomy of graptolites (see also color figure 8). (Redrawn from several sources by Mary Persis Williams)

known as a *stolon*. Likewise, many graptolites had hundreds of thecae and made up huge colonies that once floated on the world's oceans in the early Paleozoic. Some were apparently dangling down from driftwood, and others were preserved with some sort of bubble-like float that kept them hanging down from the ocean surface (fig. 18.3).

    Their habit of floating on the ocean surface explains why they are found in any kind of marine rock from deepwater shales to shallow marine lime-stones. They didn't live on the seafloor but floated above it everywhere, and when they died, they sank to the seafloor and were preserved. Deepwater shales were often formed in quiet, oxygen-poor waters with no other fossils

A typical graptolite colony of *Diplograptus* and *Dictyonema* as they might have looked floating on the Ordovician sea surface (see also color figure 8). (Illustration by Mary Persis Williams)

and no scavengers, so only graptolites sinking down from above are preserved there.

Once you look closely at the graptolite fossils, most of them are pretty easy to identify. The earliest forms from the Late Cambrian were bushy branching fossils known as "dendroids" or the genus *Dictyonema* (fig. 18.4A). This form persisted until the end of the Paleozoic, and some were attached to the hard surfaces on the seafloor. However, by the Ordovician, graptolites had become open-ocean drifters and began to simplify their structures into a few stipes on each rhabdosome. Typical Ordovician fossils include *Diplograptus* and *Didymograptus* (fig. 18.4B), whose name describes the double rows of thecae on both sides of the stipe (*diplo-* and *didymo*—both mean "double" in Greek). Another Ordovician fossil, *Phyllograptus*, had four leaf-shaped stipes that looked like the vanes on the back end of a dart (fig. 18.4C).

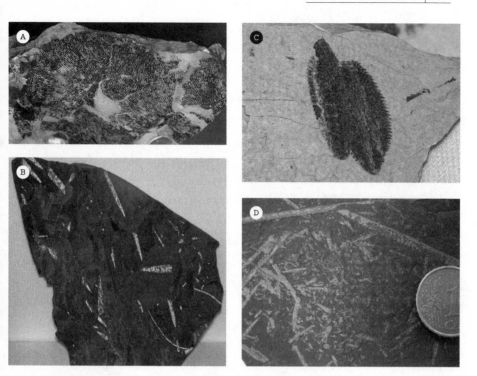

Figure 18.4 ▲

Some distinctive graptolites: (*A*) densely bushy branching primitive forms called *Dictyonema*; (*B*) double-sided Ordovician graptolite *Didymograptus*; (*C*) leaf-shaped Ordovician graptolite *Phyllograptus*; and (*D*) *Monograptus*, a graptolite with a single row of thecae on each branch that is an important Silurian-Devonian index fossil. (Courtesy of Wikimedia Commons)

By the Late Ordovician and especially in the Silurian, the most common graptolites were even more simplified. They are called *Monograptus* (fig. 18.4D) because they have just a single row of cups on each stipe (*mono* means "one" in Greek). Monograptids diversified into many different shapes in the Silurian, including corkscrew-like rhabdosomes and structures with a large fan of individual stipes flaring out. They are so useful for telling time that they define the stages of the Silurian worldwide. In addition, *Monograptus uniformis* is the official indicator of the Silurian-Devonian boundary. By the Middle Devonian, most graptolites had vanished, with only the bushy dendroid forms like *Dictyonema* surviving into the Carboniferous, and then they vanished forever. We do not know for sure why they became extinct. But the Devonian was a time of a great radiation of jawed fishes, and they would have found the tiny branching creatures floating on the surface easy prey.

Graptolites may not be as spectacular as dinosaurs or trilobites, but they are very important in telling time in the early Paleozoic and in helping us reconstruct the positions of ancient continents and oceans. In addition, they are not a difficult group to master. About 250 genera are recognized worldwide, and most of their evolution can be summarized on a single wall chart by Churkin and Carter that was published by the Geological Society of America in 1972, and it is still used by paleontologists today.

PHYLUM CHORDATA

Everyone loves dinosaurs, and most of us have at least some interest in how vertebrates evolved because we humans are vertebrates (fig. 19.1). However, most vertebrate fossils are extremely rare, and they are not easy to collect or own. The key fossils that tell the story of how vertebrates evolved from one group to another are all in museums, and paleontologists as well as the casual collector can see many of them on display, and sometimes purchase replicas.

My approach in this chapter is not to tell the entire story of vertebrate evolution, which is covered in many other books, including my books *Bringing Fossils to Life: An Introduction to Paleobiology*, *The Story of Life in 25 Fossils*, and *Evolution: What the Fossils Say and Why It Matters*. Instead, I will briefly describe the major groups that occur in important fossil localities and explain how they can be identified.

JAWLESS FISH

The earliest vertebrates did not have jaws, but they did have bony armor surrounding their bodies that was held up by an internal skeleton made of cartilage (fig. 19.2A). During the Silurian and especially the Devonian, there was a huge evolutionary radiation of these jawless fish (often called by the wastebasket name "agnathans"). These included the bottom-dwelling *osteostracans* or "ostracoderms," which had a broad flat-bottomed horseshoe-shaped head shield with upward facing eyes,

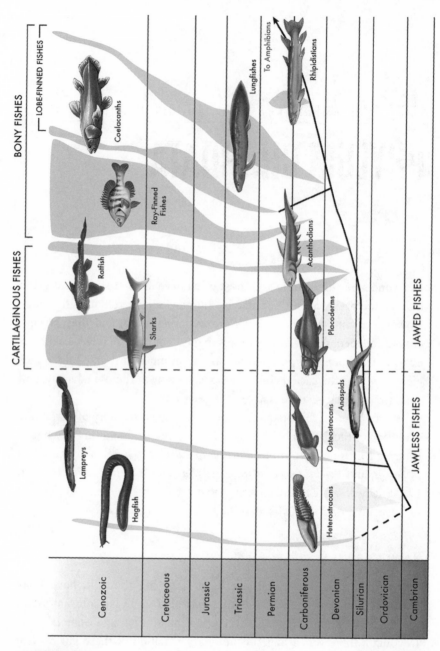

Figure 19.1 ▲

Evolutionary radiation of the vertebrates. The first branch point is the jawless fish, followed by the sharks, and then the extinct placoderms. To the left of that are the earliest relatives of advanced fish (the acanthodians), then the huge evolutionary radiation of bony fish. The remaining branches are amphibians, reptiles (with birds branching from dinosaurs), and mammals. (Illustration by Mary Persis Williams)

and a tail like that of a shark, with the lobe bent upward (fig. 19.2B); the tadpole-shaped *heterostracans*, which had a large head and body shield and a tail with the lobe pointed downward (fig. 19.2C); and the *anaspids* (fig. 19.2D) and the *thelodonts*, which were small fish with a slit-like mouth,

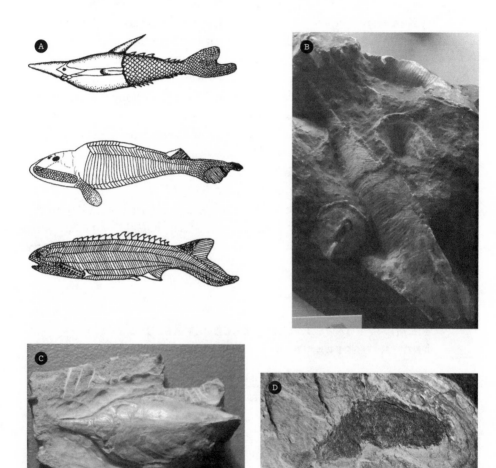

Figure 19.2 ▲

(*A*) Three typical examples of jawless fossil fish: the heterostracan *Pteraspis* (*top*), which had a streamlined head shield; the osteostracan *Hemicyclapsis* (*middle*), with the large horseshoe-shaped, flat-bottomed head shield; and the scaly *Birkenia* (*bottom*), with the slit-like mouth, downturned tail fin, and numerous spines. (*B*) Nearly complete specimens of *Hemicyclaspis*. (*C*; color figure 9) Head shield of *Pteraspis*. (*D*) Nearly complete fossil of *Birkenia*. ([*A*] Redrawn from several sources by Mary Persis Williams; [*B–D*] courtesy of Wikimedia Commons) a downward-pointing lobe on the tail, and were fully covered by a sort of chain mail of small bony plates. All of these Devonian jawless fish vanished in the Late Devonian extinction event.

The only survivors of the radiation of jawless fish are two groups of extremely specialized jawless vertebrates: lampreys and hagfish. The lampreys are adapted to latching onto the side of a fish and sucking on them as parasites. The hagfish lives in the bottom muds, slurping up worms, or eating a fish carcass from the inside. Neither of these fish is typical of their jawless ancestors. They were adapted for their peculiar scavenging lifestyle and have completely lost their bony armor; all that remains is their cartilage skeleton.

Fossils of jawless fish are extremely rare, but certain localities have famous fish beds where these fish are occasionally found by collectors. These sites include the famous fish beds of the Old Red Sandstone in England and Scotland, the Devonian beds of the Catskill Mountains of New York, Devonian beds in Scaumanec (or Escuminac) Bay in the Gaspé Peninsula in Quebec, the Cleveland Shale in Ohio, and a few other Devonian sites. Fossils of jawless fish are on display in many natural history museums, although their small size and incomplete preservation makes them less impressive to see than other fish fossils.

PRIMITIVE JAWED FISH: PLACODERMS, SHARKS, AND THEIR RELATIVES

The evolution of jaws was an important breakthrough in vertebrate evolution. Jawless fish are severely limited by their simple sucker-like mouths. They could only feed on detritus on the sea floor, filter-feed in the plankton, or become scavengers like the hagfish and lamprey. Jaws made it possible for fish to feed on many different types of food (especially larger prey) and to chop it into digestible pieces, but jaws are used by vertebrates in many other ways as well. For example, jaws are used for digging holes, carrying pebbles or vegetation to build nests, grasping mates during courtship or copulation, carrying their young around, and making sounds or speech.

Several groups of early jawed vertebrates are found in the fossil record. The earliest were an extinct group of highly armored fish known as placoderms. Their body was held up by cartilage, and large bony plates covered their head and the front of their bodies. Some of them, such as the

Figure 19.3 ▲

Examples of fossil placoderms: (*A*; color figure 10) head and body shield of the giant arthrodire *Dunkleosteus*; (*B*) reconstruction of *Dunkleosteus*; (*C*) head and body shield and armored appendage of the small, bottom-dwelling antiarch *Bothriolepis*; and (*D*) a reconstruction of *Bothriolepis* in life. (Courtesy of Wikimedia Commons)

Late Devonian monster predator *Dunkleosteus* (originally known as *Dinichthys*), were 33 feet (10 meters) long, by far the largest predators on Earth at that time (fig. 19.3A–B). The antiarchs and other smaller fish were armored all the way down to the base of the tail, and their appendages were also armored and resemble crab legs (fig. 19.3C–D). Still others were shaped like rays, or like the Port Jackson shark. The radiation of jawed placoderms dominated the Devonian seas and were particularly abundant in famous Upper Devonian fish beds such as the Old Red Sandstone in Great Britain or the Cleveland Shale in Ohio. Most are not easy to collect, but some are on display in museums around world.

The most commonly collected of primitive jawed vertebrates are the sharks, whose abundant teeth can be found in many localities around the world and are easy to collect or purchase. Sharks began to evolve in the Late Devonian side by side with the enormous radiation of jawless fish and placoderms. The latter two groups vanished during the Late Devonian extinction event, but sharks survived and continued to evolve in the late Paleozoic and Mesozoic, and they continue to thrive today. Sharks are highly diverse in the modern oceans, and a lot is known about their biology and evolution.

Sharks have an internal skeleton made of cartilage, which rarely fossilizes. However, they have some bone in the denticles in their skins (giving sharkskin its typical rough texture), in the spines on the front edge of their fins, and especially in their teeth. A shark's mouth has hundreds of teeth at any given time. Only the teeth at the front of the jaw are in active use. As these teeth become worn down or broken, they are shed and more teeth grow in behind them. This is the reason sharks leave so many teeth in the fossil record, and why they are so easy to collect and own.

The evolution of shark teeth is quite a remarkable story in itself. In the Late Devonian, primitive sharks like *Cladoselache* (fig. 19.4A–B) were up to 6 feet (2 meters) in length, and they had teeth with a simple conical cusp in the center and small pointed cuspules on each side. Unlike more advanced sharks, their pectoral fins had a rigid broad base, so they could not maneuver in the way more advanced sharks do.

In the late Paleozoic, eel-shaped xenacanth or pleuracanth sharks up to 10 feet (3 meters) long lurked in the oceans and swamps (fig. 19.4C–D). Their teeth have a pair of sharp prongs forming a V-shape. Another weird Paleozoic shark was the maker of the spiral whorls of shark teeth known as *Helicoprion* (fig. 19.4E). Recent discoveries have revealed the shape of this shark and how it carried its tooth whorl (fig. 19.4F). Later sharks have an even wider variety of shapes and sizes of teeth, and the living sharks still show tremendous variation. Some of the most popular are the huge teeth of *Carcharocles megalodon*, the gigantic great white sharks that may have reached 82 feet (25 meters) in length (fig.19.4G). Their teeth (fig. 19.4H) are particularly common in the Miocene marine rocks at places like Sharktooth Hill in California, the Calvert Cliffs in Chesapeake Bay in Maryland, or Lee Creek Mine in North Carolina. The distinctive asymmetrical triangles of mako shark (*Isurus*) teeth are also commonly found in those same beds, as

Figure 19.4 ▲

Fossil sharks: (*A*) the primitive Devonian fossil shark *Cladoselache*, from the Cleveland Shale and (*B*) a reconstruction of *Cladoselache*; (*C*) the peculiar Carboniferous-Permian xenacanth or pleuracanth sharks and (*D*) a reconstruction. (*E*) These weird tooth whorls called *Helicoprion* were long a mystery. (*F*) Modern restoration of the shark that carried the tooth whorls. (*G*) Life-sized reconstruction of the gigantic great white shark *Carcharocles megalodon* from the Miocene and (*H*) a typical *C. megalodon* tooth surrounded by teeth of *Isurus*, the mako shark. ([*A, C, E*] Courtesy of Wikimedia Commons; [*B, D*] courtesy of N. Tamura; [*F–G*] photographs by the author; [*H*] courtesy of R. Irmis)

Figure 19.4 ▲
(continued)

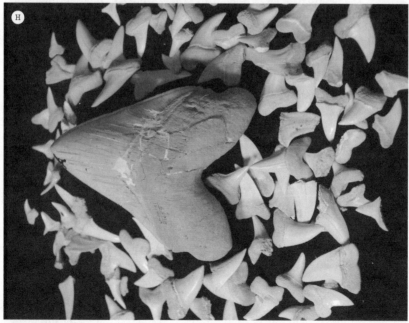

Figure 19.4 ▲
(*continued*)

well as the pavement teeth of skates and rays. Collecting fossil shark teeth is popular in many parts of the world, and many different kinds of shark teeth are available for sale on the commercial market.

BONY FISH: OSTEICHTHYES

The skeletons of sharks, placoderms, and jawless fish are mostly made of cartilage; they have bone only in their external armor, their teeth, or spines. The other main branch of fish is the bony fish, or Osteichthyes. Their skeleton is mostly made of bone, and they have additional bones on the surface of their body, on their skull, and elsewhere.

One branch of the osteichthyans is the lobe-finned fishes, or *sarcopterygians*. They have fleshy and muscular fins and have just a few robust bones that match the bones in our arms and legs. Lobe-finned fish include the fossil and living lungfish, the coelacanths, and an extinct wastebasket group called the "rhipidistians," which gave rise to amphibians. Nearly all of these fish are rare and hard to collect.

The other main branch of bony fish is the *actinopterygians*, or ray-finned fish, which includes fish whose fins are supported by many parallel rays of bone. A Devonian group called the palaeoniscoids have the most primitive structures among ray-finned fish (fig. 19.5A–B). The next most advanced fishes in this lineage are represented by the sturgeon, the paddlefish, the bichir, and their extinct relatives from the late Paleozoic, which used to be lumped into a wastebasket group called the "chondrosteans." These fish have dense areas of bone in their skulls but have a primitive jaw mechanism and primitive fins (especially the tail fin). During the Mesozoic, another great evolutionary radiation of more advanced fish, known by the wastebasket group "holosteans," were the dominant fish on Earth, but today only a few survive, including the gar fish (fig. 19.5C) and *Amia*, the bowfin. All of these archaic fish are collected in important Paleozoic and Mesozoic fish localities, and some are also sold on the commercial market as well.

By far, the largest group of fish is the advanced ray-finned fish, or the *teleosts*. There are more than 20,000 living species of teleosts; they are as diverse as all other fish, amphibians, reptiles, mammals, and birds put together. In fact, nearly every familiar fish you find in an aquarium or fish market or anywhere else (except for sharks and the fish just mentioned) is a teleost. Although we are mammal chauvinists and like to call the Cenozoic

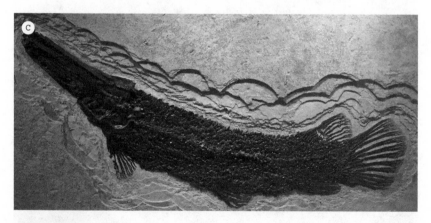

Figure 19.5 ▲

Examples of well-known fossil bony fish: (*A*) the primitive bony fish from the Permian, *Palaeoniscus* and (*B*) a reconstruction of the palaeoniscoid fish *Cheirolepis*; (*C*) a large gar *Lepisosteus*, one of the most primitive living bony fish, from the Green River Formation. (*D*) This specimen of the gigantic ichthyodectiform *Xiphactinus* from the Cretaceous chalk beds of western Kansas has swallowed a smaller fish. (*E*) The herring *Knightia* is the most common fish from the Eocene Green River Formation; and (*F*) the larger herring relative *Diplomystus* is the second most common fish in the Green River Formation. ([*A, C, E*] Courtesy of Wikimedia Commons; [*B*] courtesy of N. Tamura; [*D, F*] photograph by the author)

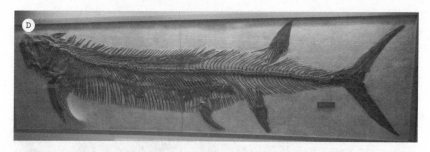

Figure 19.5 ▲
(continued)

the Age of Mammals, it could really be called the Age of Fish. In fact, a lot more species of teleosts were evolving in the Cretaceous or Cenozoic than there were dinosaurs or mammals. Among the most primitive teleosts were the giant, impressive fish called *Xiphactinus* (formerly *Portheus*), from the Cretaceous seaways of western Kansas; you will find them on display in many places in the world (fig. 19.5D).

Nearly any fish fossil that is easy to collect in the United States or that is available on the open market is a teleost as well. This includes the fossils found in huge deposits of Eocene lake sediments of the Green River Formation in Wyoming, Utah, and Colorado, which have produced thousands of beautiful fish specimens, especially the common small fish *Knightia* or the larger *Diplomystus*, which can be found in all the commercial markets (fig. 19.5E–F). A number of other localities around the world produce complete, articulated fish specimens like those of the Green River Shale. Fossil fish can be found in just about any exposure of the Green River Shale, although the best collecting is in commercial fish quarries such as the Ulrich operation, which tends to be clustered around Fossil Butte National Monument in Kemmerer, Wyoming. Other legendary localities for complete teleost fish include the Eocene beds of Monte Bolca in Italy and Messel in Germany. In collecting, however, it is much more typical to find only a partial fish, or a bunch of isolated fish bones. A good fish paleontologist can identify most familiar fishes from just a few isolated bones.

CLASS AMPHIBIA: AMPHIBIANS AND THEIR RELATIVES

During the Late Devonian, one group of lobe-finned fishes known as "rhipidistians" gave rise to the first great radiation of land vertebrates, the four-legged animals, or tetrapods, informally known as the "amphibians." We now have many fossils that show this transition from lobe-finned fish to tetrapod, including *Tiktaalik* (see chapter 4) and *Ichthyostega* and *Acanthostega*. These more advanced transitional fossils still have fish-like features, such as gill slits and a large fin on the tail, and they have sense organs adapted for water and not land. Although their lobed fins were modified into limbs for crawling, analyses show they probably didn't crawl on land much but used their limbs for moving along the water bottom, as modern newts do.

By the late Paleozoic, these primitive tetrapods had evolved into three major groups that were among the largest land animals, especially during the Carboniferous and Permian. They were first discovered in Upper Carboniferous mines that were extracting the coal made in ancient swamps where they once lived. In some mines, skulls were found embedded in the mine walls, and others were exposed when coal-rich shales were split open. The best known of these tetrapods come from the Lower Permian red beds around Seymour, Texas, where they can be collected today. One group of tetrapod was the long, flat-bodied crocodile-like *temnospondyls* (once called "labyrinthodonts"). The biggest ones included *Eryops*, which reached 6 feet (2 meters) in length, with a big flat skull that was one-fifth its body length (fig.19.6A). An even bigger temnospondyl from the Lower Permian red beds of Texas was *Edops*, which is known only from a skull, but that skull is quite a bit larger than *Eryops*. In the Triassic, the temnospondyls were in decline, but huge flat-bodied metoposaurs like *Buettneria* (fig. 19.6B–C) were still common in the Petrified Forest National Park and in the Painted Desert of Arizona.

A second radiation of tetrapods in the late Paleozoic was the *lepospondyls*, which tended to be smaller and more like salamanders. One group, the microsaurs, was extremely lizard-like, with a deep skull, a cylindrical body, and relatively tiny limbs. Another group, the aistopods, secondarily evolved a limbless snake-like body, convergent on many other animals (including not only snakes but also apodan amphibians and amphisbaenid reptiles) that have lost their legs and become snake-like. The most odd-looking lepospondyls from the Lower Permian red beds are nectrideans like *Diplocaulus*, which had a huge boomerang-shaped head on a salamander-like body (fig. 19.6D).

The third major group of late Paleozoic tetrapods was the *anthracosaurs* ("coal lizards" in Greek because many early specimens were found in coal deposits). This wastebasket group is actually more closely related to reptiles than it is to amphibians. Like reptiles, they had a deep skull with

Figure 19.6 ▶

Examples of well-known extinct amphibians: (*A*) the giant Permian temnospondyl *Eryops*; (*B*) the flat-bodied Triassic temnospondyl *Buettneria* and (*C*) a reconstruction of *Buettneria* in life; (*D*) the boomerang-headed lepospondyl *Diplocaulus*; (*E*) the reptile-like "anthracosaur" *Seymouria*; (*F*) the transitional fossil between frogs and salamanders, *Gerobatrachus*, nicknamed the "Frogamander"; and (*G*) the Permian fossil closest to the root of modern amphibians known as *Cacops*. ([*A–E*] Photographs by the author; [*F*] courtesy of J. Anderson; [*G*] courtesy of Wikimedia Commons)

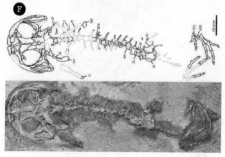

Figure 19.6 ▲
(*continued*)

a short snout (rather than the flat skull with the long snout in temnospondyls), large eyes, strong legs, and a relatively short body and tail, suggesting that they were much less aquatic and more terrestrial than the other two groups. They show many anatomical features of reptiles as well, suggesting that they are transitional forms between amphibians and reptiles. Some, like *Seymouria* from the Lower Permian red beds near Seymour, Texas, are remarkably reptilian in all but a few features (fig. 19.6E).

All three of these groups of archaic tetrapods were extinct by the end of the Triassic. However, the Lower Permian red beds of Texas also yield fossils such as *Gerobatrachus hottoni*, the "Frogamander," a fossil with the head and broad snout of a frog but the body of a salamander (fig. 19.6F). By the Triassic, numerous transitional fossils (such as *Triadobatrachus*) to frogs had a basically frog-like build but still did not have the specialized spine and hips seen in modern frogs, nor were their legs as long and adapted for leaping. These fossils, as well as all living frogs and salamanders, have a number of distinctive anatomical features (such as teeth on pedestals with a distinctive base) that link them to a group of Permian temnospondyls like *Cacops* (fig. 19.6G) and *Doleserpeton*. From these early Mesozoic origins, frogs and salamanders have undergone a huge evolutionary radiation, and 4,000 species are still alive today. Frog and salamander fossils are not common, but complete frog fossils have been collected from the Green River Shale of Wyoming.

CLASS REPTILIA: REPTILES

We are all familiar with the living reptiles, including turtles, snakes, lizards, and crocodilians, but the fossil record of reptiles includes many more groups, especially dinosaurs, plus the flying reptiles (pterosaurs), the marine reptiles such as the dolphin-like ichthyosaurs and the paddling plesiosaurs, and many others (fig. 19.7). Most extinct reptiles, such as dinosaurs, were quite rare compared to invertebrate fossils, and good fossil specimens are normally collected by professional researchers for museums. But many other fossil reptiles are relatively easy to collect, and many people have done so or bought them from rock shops and other dealers.

The living reptiles fall into several groups, which have many more fossil relatives. The most primitive living reptiles are turtles and tortoises (Anapsida), and they have many extinct relatives in the late Paleozoic and Triassic. The remaining reptiles are now lumped into a broadly defined group called the Diaspida. One branch, the Euryapsida, includes the extinct marine reptiles such as the plesiosaurs and ichthyosaurs. The other branch (the Sauria) is divided into two main groups: the lizards and snakes (Lepidosauria) and their relatives, and the Archosauria, which includes crocodilians, dinosaurs, birds, and many extinct groups.

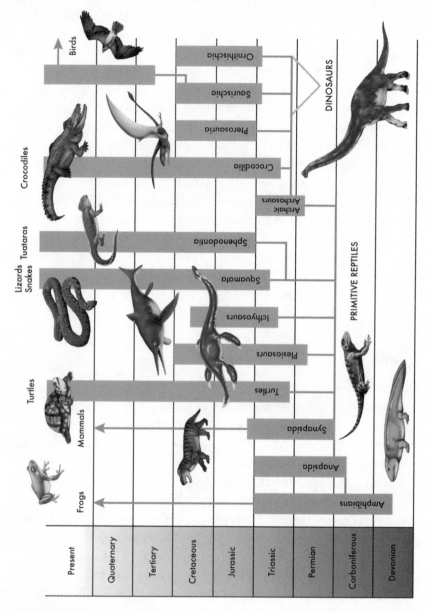

Figure 19.7 ▲

Family tree of reptiles and other amniotes. The anapsids, the most primitive amniotes known (such as the Carboniferous fossils *Westlothiana* and *Hylonomus*), gave rise to two branches: the synapsids and their mammalian descendants and the Reptilia. The most primitive branch of reptiles is the turtles, and the dolphin-like ichthyosaurs and the paddling plesiosaurs are both major groups of Mesozoic marine reptiles. The Lepidosauria include the living sphenodontids (the tuatara and its ancestors) and the squamates, or the lizards and snakes. The Archosauria include the archaic archosaurs (formerly called "thecodonts"), the crocodilian branch, the flying pterosaurs, and the dinosaurs. One branch, the Saurischia, includes the giant sauropods and the predatory theropods, which gave rise to the birds. The ornithischian dinosaurs include mostly herbivorous forms: duckbills, horned dinosaurs, stegosaurs, ankylosaurs, and their kin. (Illustration by Mary Persis Williams)

ANAPSIDA: TURTLES AND THEIR RELATIVES

The anapsids include a variety of strange reptiles of the late Paleozoic. The huge pariesaurs were heavily built, thick-limbed hippo-sized herbivores whose dense bony skulls were covered with warts and knobs of bone. Others, such as the procolophonids and captorhinids, were built like medium-sized lizards, but the details of their skulls were very turtle-like. The turtles and tortoises have a very impressive fossil record because they tend to have heavy dense bones and their shells are easily fossilized. The oldest turtle relatives from the Permian include *Eunotosaurus* from 260 million years ago, which already had broadly flared ribs in the back that were precursors to their back shell (carapace), as well as many other turtle-like features of the skeleton. By 240 million years ago, fossils like *Pappochelys* ("grandfather turtle" in Greek) had broad flattened ribs on their back and also broad ribs in the belly region called gastralia, which were the precursor to the belly shell (plastron). Even more of a transitional fossil is *Odontochelys semitestacea*, the "turtle on the half shell" from the Triassic of China (fig. 19.8A–B). Its name literally means "half-shelled toothed turtle" in Greek; it had a shell on its belly but still had only the broad flared ribs on its back, not a carapace. In addition, it was the last turtle known to still have teeth; all later turtles had a toothless beak. By 215 million years ago, in the Late Triassic, we see fossil turtles like *Proganochelys* (fig. 19.8C), which had a complete carapace, but no teeth. It also had many other advanced turtle features. However, unlike any living group of turtles, *Proganochelys* could not pull its neck and head into its shell; instead, its head was heavily armored. By the Cretaceous, enormous sea turtles like *Archelon*, from the chalk beds of western Kansas (fig. 19.8D), were more than 12 feet (3.5 meters) long. In the Cenozoic, the largest of the enormous freshwater turtles was *Stupendemys* from the Miocene of Brazil (fig. 19.8E). Its shell alone was 11 feet (3.3 meters) long.

During the Mesozoic, the major group of living turtles evolved, and they tended to leave an excellent fossil record because their shell is so thick and durable. In many dinosaur beds around the world, the first fossils found are broken scraps of turtle shell. Although these scraps are hard to identify, they are a good indicator that you're collecting in the right place. In some fossil beds, such as the Big Badlands of South Dakota, fragments of shell of the tortoise *Stylemys nebrascensis* (fig. 19.8F) are the most common fossils found, and occasionally a complete turtle shell is found.

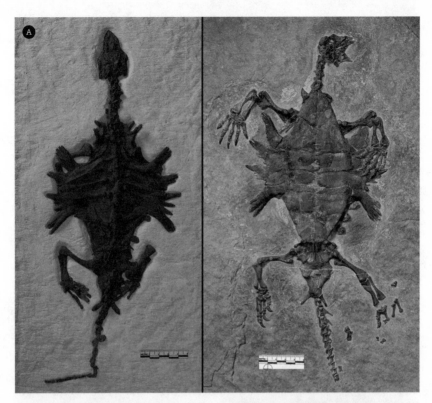

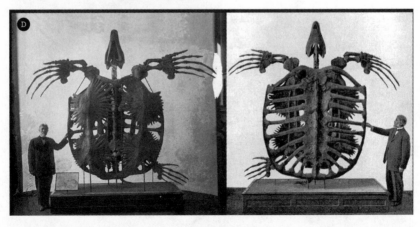

Figure 19.8 ◄ ▲

Examples of well-known fossil turtles. (*A*) *Odontochelys semitestacea*, the Permian turtle with a shell on its belly but not on its back, and still retaining teeth, and (B) a reconstruction showing shell placement. (*C*) *Proganochelys*, a Triassic turtle with both shells but no teeth and a nonretractable head covered by armor. (*D*) The enormous Cretaceous sea turtle *Archelon*, the largest turtle that ever swam the oceans. (*E*) The shell of the gigantic Miocene freshwater turtle *Stupendemys*, the largest turtle on the land. (*F*) The common tortoise of the early Oligocene of the Big Badlands of South Dakota, *Stylemys nebrascensis*. ([*A–B*] Courtesy of Wikimedia Commons; [*C–F*] courtesy of Wikimedia Commons)

EURYAPSIDA: ICHTHYOSAURS AND PLESIOSAURS

During the Mesozoic, the top predators in the oceans included a variety of reptiles. There were enormous sea turtles like *Archelon*, but the biggest predators were all euryapsids. Two main groups fought for control of the seas: ichthyosaurs and plesiosaurs.

The most specialized were the dolphin-like ichthyosaurs, which had a highly streamlined fish-like body (fig. 19.9A). Their hands and feet were completely modified to flippers, and they had a large dorsal fin, a tail fin, and a long narrow snout full of conical teeth for catching aquatic prey like fish, ammonites, and squid. Most of them had large eyes for seeing in dim murky water. The largest included the huge 49-foot-long whale-sized *Shonisaurus* (fig. 19.9B). This amazing creature can be seen at Berlin-Ichthyosaur State Park near Gabbs, Nevada, and it also is on display in the Nevada State Museum in Las Vegas; it is the official Nevada State Fossil. Many museums display extraordinary complete articulated skeletons of ichthyosaurs that even show the dark film of the outline of their bodies (see fig. 2.4B), typically from the Jurassic Holzmaden Shale of Germany (fig. 19.9C). Ichthyosaurs had short paddles and fish-like bodies, so they could not crawl onto land to lay their eggs as sea turtles do. They must have given live birth in the ocean, as dolphins and whales do. One of the Holzmaden ichthyosaurs was fossilized in the process of giving birth (fig. 19.9D); the baby can be seen emerging from the birth canal.

The largest marine reptiles, however, were the plesiosaurs (fig. 19.10A). They had massive bodies with huge paddles on all four limbs to slowly but steadily row through the water. One group (the elasmosaurs) had a long snake-like neck that allowed them to reach prey (fish and squid) far from their rather slow-moving bodies (fig. 19.10B). The largest of these elasmosaurs were up to 33 feet (10 meters) long. But even bigger were the pliosaurs, which had a short neck but a massive head with long jaws. The biggest of these was *Kronosaurus* from the Cretaceous of Australia, which reached 42 feet (12.8 meters) in length (fig. 19.10C). Popular media has reported that there were even larger plesiosaurs, such as *Liopleurodon*, which reached 82 feet (25 meters) in length. However, these claims were made from examining incomplete specimens, and there is no strong evidence to back up this assertion. Plesiosaurs have been found in many parts

Figure 19.9 ▲

Ichthyosaurs: (A) reconstruction of a typical *Ichthyosaurus* from the Jurassic; (B) the largest known ichthyosaur, *Shonisaurus*, from the Triassic Berlin-Ichthyosaur State Park in Nevada; and (C) two large ichthyosaurs in the Natural History Museum in London. (D) Some of the Holzmaden ichthyosaurs were buried and fossilized at the very moment that they were giving birth. The baby ichthyosaur can be seen emerging tail first from the birth canal. ([A] Courtesy of Wikimedia Commons; [B–D] photographs by the author)

of the world, especially in the Jurassic beds of Europe (fig. 19.10D), and they are particularly common in the Cretaceous beds of the Great Plains. Many elasmosaurs have been found in the chalk beds of Gove County in western Kansas, and others are still found there today.

Plesiosaurs. (*A*) An 1863 reconstruction of an ichthyosaur battling a plesiosaur in the Jurassic seas of England. (*B*) *Elasmosaurus* had the longest neck of any plesiosaur. (*C*) Pliosaurs were short-necked but long-snouted plesiosaurs. This is *Kronosaurus*, the largest plesiosaur ever found. (*D*) Typical Jurassic plesiosaurs had intermediate neck lengths and smaller heads, such as this *Rhomaleosaurus* from England. ([*A–B*] Courtesy of Wikimedia Commons; [*C*] courtesy of Ernst Mayr Library, Harvard University; [*D*] photograph by the author)

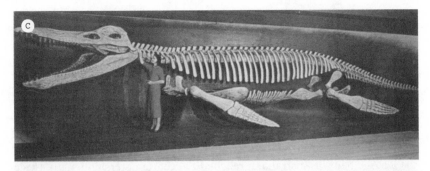

Figure 19.10 ▲
(*continued*)

LEPIDOSAURIA: LIZARDS AND SNAKES

Lizards and snakes are the most common and familiar reptiles we see today. More than 6,000 species of lizards and 2,500 species of snakes (compared to only 250 species of turtles and about 25 species of crocodilians) are alive today. The living lepidosaurs are recognized by a number of unique anatomical features, including their forked tongue used to taste the scent in the air, their distinctive scales made of two kinds of keratin protein, and numerous other features. Except for the lizard-like tuatara (a sphenodontid) from the islands of New Zealand, all the remaining lepidosaurs are *squamates*, with many distinctive anatomical features, including a skull made of bony struts so it is highly flexible and a jaw that can open very wide to eat large prey.

The oldest unquestioned lizard fossils are from the Late Jurassic. They subsequently diversified into dozens of families with over 6,000 species during the Cretaceous and the Cenozoic. Complete lizard fossils are relatively rare and tend to be found only in extraordinary fossil deposits, such as the Eocene Green River Shale of Wyoming (fig. 19.11A) or the Messel beds in Germany. However, if you collect tiny fossils in Mesozoic and Cenozoic beds, individual small bones of lizards can be quite common. Some lizards were truly impressive, such as the gigantic monitor lizard *Megalania* from the Ice Age beds of Australia. It reached 20 feet (6 meters) in length, twice the size of its close living relatives, the Komodo dragon (fig. 19.11B).

The monitor lizards had an important descendant: the marine reptiles known as mosasaurs (fig. 19.11C–D). These were just like Komodo dragons in their basic anatomy, but their hands and feet were modified into flippers, and their long flat tail helped them swim. Mosasaurs became famous when they were featured in the movie *Jurassic World*. The creature that jumps out of the water in that movie to eat a shark is much larger than any mosasaur ever known. Nevertheless, mosasaurs were one of the dominant marine predators in the Late Cretaceous. Many have been found in the inland sea deposits of Kansas and South Dakota, and their fossils (especially teeth) are still collected there today.

During the Cretaceous, one group of lizards (probably the monitor lizards, or Varanidae) became specialized for burrowing and lost their limbs to become snakes. Numerous fossils of Cretaceous snakes with tiny vestigial front and hind limbs are now known, showing how the transition took place (fig. 19.11E). Complete snake fossils are even more rare, and they only occur in a few unusual deposits such as the Eocene Green River Shale (fig. 19.11F). Their individual vertebrae are often found along with other tiny fossils in Cretaceous and Cenozoic beds in many places. Some of them reached enormous size, such as the Paleocene anaconda from Colombia known as *Titanoboa*, which was as long as a school bus (fig. 19.11G).

ARCHOSAURIA: CROCODILIANS, PTEROSAURS, DINOSAURS, BIRDS, AND THEIR RELATIVES

Based on many lines of both anatomical and molecular evidence, crocodilians and birds are closely related among all living vertebrates. Together they define a group known as the *archosaurs* ("ruling reptiles" in Greek), which also includes dinosaurs, pterosaurs, and many other extinct groups.

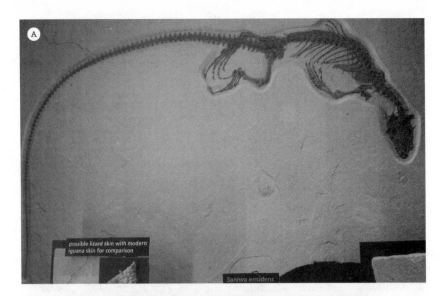

Figure 19.11 ▲

Lizard and snake fossils and their kin. (*A*) The lizard *Saniwa* from the Eocene Green River Shale. (*B*) The giant Komodo dragon *Megalania* from the Ice Age beds of Australia. (*C*) The enormous mosasaur *Tylosaurus*, more than 43 feet (13 meters) long. (*D*) Reconstruction of a mosasaur swimming in the Cretaceous seas. (*E*) The Cretaceous fossil snake *Tetrapodophis*, which still had four tiny vestigial legs. (*F*) The Eocene Green River boa constrictor *Boavus*. (*G*; color figure 11) Reconstruction of the school-bus-sized Paleocene anaconda *Titanoboa*. ([*A–C, E*] Courtesy of Wikimedia Commons; [*D, F–G*] photographs by the author)

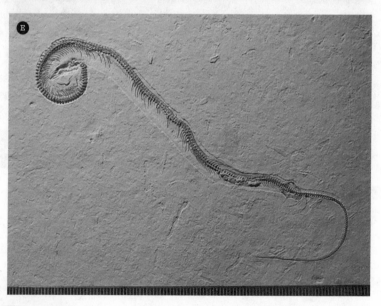

Figure 19.11 ▲

(*continued*)

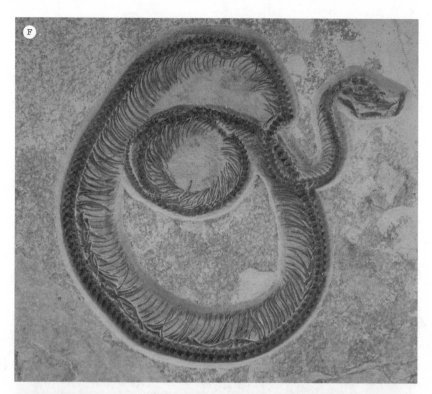

Figure 19.11 ▲

(*continued*)

Many unique anatomical features help us recognize all archosaurs, including distinctive features of the skull, lower jaw, braincase, and especially their tendency to have an upright or semi-upright posture, with their limbs beneath their bodies, rather than sprawling on their bellies with their limbs splayed out to the side as lizards do. The living crocodilians and birds have many other anatomical features that do not fossilize, such as the structure of their three-chambered heart, the muscular diaphragm to assist breathing, and a muscular gizzard in their digestive tract to grind up food (very few archosaurs have teeth or jaws that allow much chewing).

During the Triassic, a great radiation of primitive archosaurs dominated all the terrestrial habitats on Earth. Some were herbivores, such as the pig-like rhynchosaurs, which had hooked beaks and short stubby legs (fig. 19.12A). The aetosaurs had backs covered in thick plates of armor, often with spines sticking out of their sides and shoulders, and an upturned snout (fig. 19.12B).

Most, however, were carnivorous reptiles. The erythrosuchids ("bloody crocodiles" in Greek), also known as the rauisuchids, were the largest predators, with long quadrupedal bodies and deep skulls armored with sharp recurved teeth (fig. 19.12B). The phytosaurs looked very much like large crocodiles, except that they had their nostrils on the top of their heads, not on the tip of the snout as in all crocodilians (fig. 19.12C). The earliest crocodilians can be found in the Late Triassic as well, but they were small, skinny, long-legged creatures with short snouts. They did not begin to become big aquatic long-snouted predators until the phytosaurs vanished and vacated that ecological niche.

During the Jurassic and Cretaceous, there were many now extinct lineages of crocodilians. Some were completely marine with a tail fin (geosaurs), and *Sebecus* had a tall narrow snout like *T. rex*. Then there were the huge Cretaceous crocodilians *Deinosuchus* ("terrible crocodile" in Greek) and *Sarcosuchus* ("flesh crocodile" in Greek), which were 50 feet (15 meters) long, with a skull over 6 feet (2 meters) long. They were large enough to eat small dinosaurs (fig. 19.12D). In the Miocene of South America, the giant caiman alligator *Purussaurus* was almost as large (fig. 19.12E).

The crocodilian branch of the archosaurs dominated in the Triassic but were extinct by the Jurassic (except for crocodilians) because the other branch of archosaurs, the dinosaurs, arose to crowd them out. This second branch includes the flying reptiles, or pterosaurs, which first appear in

Archosaurs and crocodilians. (A) The Triassic herbivores known as rhynchosaurs had a beak and a long, low-slung body. (B) In Petrified Forest National Park, the erythrosuchids (*left*) were major predators, and their prey, which included aetosaurs (*right*), were protected with armored backs fringed with spikes. (C) The crocodile-mimic phytosaurs had a long narrow snout for catching fish, but they were unrelated to crocodilians. (D) The enormous Cretaceous crocodilian *Deinosuchus* probably preyed on small dinosaurs. (E) The Miocene caiman *Purussaurus* was almost as large. ([A] Courtesy of N. Tamura; [B–C, E] photographs by the author; [D] courtesy of the American Museum of Natural History)

Figure 19.12 ▲
(*continued*)

the Late Triassic. The earliest pterosaur, *Eudimorphodon*, had a primitive archosaurian head but fully developed wings and the ability to fly. Through the rest of the Jurassic and the Cretaceous, pterosaurs evolved into an incredible variety of forms, from small crow-sized pterosaurs like *Ptero-dactylus* (fig. 19.13A) and the vane-tailed *Rhamphorhynchus* (fig. 19.13B) to the huge *Pteranodon* (fig. 19.13C–D) with a wingspan of 22 feet (7 meters).

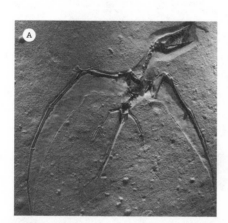
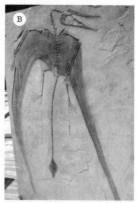
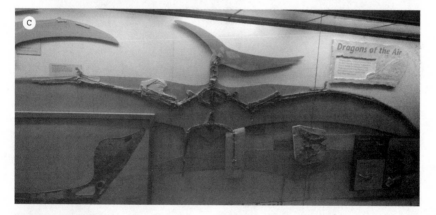

Figure 19.13 ▲

Pterosaurs were flying archosaurs closely related to dinosaurs but not members of the Dinosauria. (*A*) The robin-sized *Pterodactylus* from the Jurassic Solnhofen Limestone. (*B*) The vane-tailed *Rhamphorhynchus* from the Solnhofen. (*C*) A nearly complete skeleton of *Pteranodon*, showing their huge size. (*D*) The huge toothless crested *Pteranodon* sailed over the Cretaceous seas of Kansas. (*E*) The largest pterosaurs of all were azhdarchids, such as this giant Texas pterosaur *Quetzalcoatlus*, which was the size of a small airplane. ([*A–D*] Photographs by the author; [*E*] courtesy of Wikimedia Commons)

Figure 19.13 ▲
(*continued*)

They had a crest on the back their head and long toothless jaws, and they once soared over the seas of Kansas. The biggest of the pterosaurs was the enormous Texas pterosaur *Quetzalcoatlus*, which was the size of a small airplane (fig. 19.13E).

Dinosaurs are the most popular of all extinct prehistoric creatures, and people have read and heard lots of things about them. However, not every extinct animal is a dinosaur. Many books and packages of plastic toy dinosaurs include animals that are *not* dinosaurs. For example, saber-toothed cats and mammoths are mammals, not dinosaurs. Pterosaurs are closely related to dinosaurs, but they were not members of Dinosauria, so calling them "flying dinosaurs" is incorrect. Plesiosaurs and ichthyosaurs and other marine reptiles are not dinosaurs—in fact, there were *no* marine dinosaurs. The fin-backed creature called *Dimetrodon* is related to mammals, and it is not even a reptile. Being extinct or prehistoric does not make something a dinosaur.

So what makes a dinosaur? Dinosaurs are defined by a very specific set of anatomical features. For example, only dinosaurs have a hip joint that is an open hole and not an enclosed socket like ours (fig. 19.14). Their limbs are held completely vertical beneath their bodies, so the head of the thighbone has a right-angle bend where it inserts into the hip socket. They also walked on the tips of their toes, not on the palms of their hands or soles of their feet. In their hind legs, the joint between the leg bones and the foot occurs not between the shinbone and the first row of ankle bones (as in most vertebrates) but between the first and second row of ankle bones. When you eat the drumstick of a chicken or turkey, the cap of cartilage on the end is actually the first row of ankle bones. In dinosaurs, this first row of ankle bones also has a spur of bone that runs up the front of the shinbone. These and many more unique anatomical features are what make a creature a dinosaur.

The earliest dinosaurs were turkey-sized bipedal runners like *Eoraptor* and *Herrerasaurus* from the Late Triassic of Argentina (fig. 19.15). Soon their descendants split into two main branches: the Saurischia, or "lizard-hipped" dinosaurs, and the Ornithischia, or "bird-hipped" dinosaurs (see fig. 19.14). The Saurischia include two main groups, the long-necked long-tailed huge sauropods, such as "Brontosaurus" and *Brachiosaurus*, and the predatory theropod dinosaurs (fig. 19.16A–D). Their hip bones have the primitive lizard-like configuration, with the pubic bone in front of the hip socket pointing forward and down.

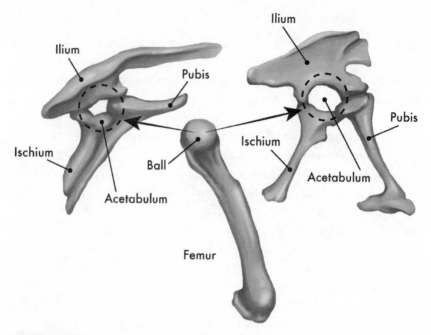

Figure 19.14 ▲

The Dinosauria are defined by having an open hole through their hip socket, into which the ball joint of the thighbone (femur) fits. The two major groups are the Saurischia (*right*), where the pubis points forward only, and the Ornithischia (*left*), where part or all of the pubis runs backward parallel to the ischium. (Illustration by Mary Persis Williams)

Figure 19.15 ▲

The most primitive dinosaurs were small, fast-running bipedal predators like *Herrerasaurus* (the larger one) and *Eoraptor* (the smaller one), from the Late Triassic of Argentina. (Courtesy of Wikimedia Commons)

Figure 19.16 ▲

Typical saurischian dinosaurs. (*A*) The small Late Triassic theropod *Coelophysis*. (*B*) *Tyranno-saurus rex*, the largest land predator in North America. (*C*) The dinosaur called "Velociraptor" in the *Jurassic Park* movies is actually *Deinonychus*; true *Velociraptor* was the size of a turkey. (*D*; color figure 12) The giant brachiosaur sauropod *Giraffatitan*, on display in Museum für Naturkunde in Berlin. ([*A–C*] Photographs by the author; [*D*] courtesy of Wikimedia Commons)

Figure 19.16 ▲
(continued)

In the Ornithischia, on the other hand, at least part of the pubic bone points backward. Ornithischians are entirely herbivorous dinosaurs, including duckbills, iguanodonts, stegosaurs, the armored ankylosaurs, the thick-headed pachycephalosaurs, and the horned and frilled ceratopsians (fig. 19.17A–D).

Figure 19.17 ▲

Typical ornithischian dinosaurs. (*A*) The most common and diverse ornithischians were duck-bill dinosaurs; this is *Edmontosaurus*. (*B*) The turtle-like armored ankylosaurs came in a variety of sizes and shapes; this one is *Gargoylesaurus*. (*C*) Stegosaurs had armored plates along their spine, tiny heads, and four spikes on their tails. (*D*) *Triceratops* and the ceratopsians had frills on the back of their neck and different combinations of horns. (Photographs by the author)

Figure 19.17 ▲
(continued)

In addition to a better understanding of how dinosaurs are interrelated and how they evolved, we also understand their biology much better. Numerous discoveries of feathers in nonbird dinosaurs have demonstrated that down feathers and body feathers occurred in every branch of the dinosaurs. If you see dinosaurs reconstructed without feathers (as in the *Jurassic Park/Jurassic World* movie), it is grossly out of date.

Much of what the public thinks about dinosaurs comes from movies and TV shows, but we must keep in mind that how dinosaurs behaved or what they sounded like or what color they were is mostly guesswork. In most cases, we have only the bones, which don't tell us much about the colors of dinosaurs or how they acted. In a few cases, feathered dinosaurs were preserved with their pigment cells, so we can tell what colors they were, but for most dinosaurs we simply don't know. The sounds of dinosaurs are pure guesswork, although a few dinosaurs, such as *Parasaurolophus*, had crests that might have produced a hooting or honking sound. Dinosaur behavior is even harder to determine scientifically. A few trackways have been discovered that show how fast dinosaurs could move and prove that they did not drag their tails but carried them straight out behind them. Dinosaur fossils occasionally show damage from combat between individuals, or tooth marks showing a predator or scavenger eating a carcass. However, the popular scenario of *Triceratops* fighting a *Tyrannosaurus rex* is still not proven to have happened (fig. 19.18). These two dinosaurs did live at the same place and time, so they probably did fight. But we have yet to find any *T. rex* bite marks on a *Triceratops* indicative of a battle, although bite marks on the body of *Triceratops* do suggest that they have been scavenged.

In addition, there has long been a controversy about dinosaur physiology. Since the 1970s, a number of paleontologists have argued that dinosaurs were "warm blooded." In other words, like birds and mammals, dinosaurs had *endothermy*, or body heat generated by their internal metabolism rather than from the environment. Since the late 1970s, most of the evidence for endothermy in all dinosaurs has been shown to be ambiguous or inconclusive. Most paleontologists now agree that the smaller dinosaurs and pterosaurs were very active animals that ran and flew, so they probably were endothermic. These animals have a large surface area relative to their small volume, and they would lose heat at a very rapid rate unless they had some kind of insulating covering, such as fur or feathers.

Figure 19.18 ▲

The skeletons of *Tyrannosaurus rex* and *Triceratops*, posed for battle. (Photograph by the author)

The largest dinosaurs have the opposite problem. As body size increases, the surface area increases only as a square, but the volume increases as a cube. So large animals have too small a surface area for their volume. For example, large living terrestrial animals (such as elephants) have a problem getting rid of heat. Elephants manage by flapping their ears (which are full of blood vessels and are primarily used for dumping heat) and by bathing frequently. Camels can allow their body temperature to fluctuate throughout the day. They are cold in the morning, take a long time to warm up in the desert, and then slowly cool down at night. This kind of thermal inertia means that large dinosaurs probably had a constant body temperature by virtue of their body size alone (*inertial homeothermy* or *gigantothermy*). They would not have needed any special regulatory mechanism or internal source of heat in the warm climates of the Mesozoic. Indeed, if they had been endothermic, they would have had problems getting rid of their excess body heat.

Dinosaur fossils are glamorous and are the highlights of exhibits for many museums today, but they are not easy to collect or to possess. Most dinosaur fossils are found on public land, and you need the proper permits to collect these fossils. Government land management agencies need to know that you are qualified to do the collecting properly before a permit is granted. Fossils are also found on private ranch land, and you need the permission of the owner to collect on this land. Poaching dinosaur fossils and selling them for obscenely high prices has become an issue today, and many ranchers are unwilling to let collectors on their land without receiving a large payment up front.

Fragmentary scraps of dinosaur bone can be found quite easily or bought on the commercial market because they have little scientific value. If you prospect long enough in some of the classic dinosaur beds of the Rocky Mountains, you will find them (along with lots of turtle shell fragments, crocodile teeth, and other scraps). Some of these classic dinosaur beds include the Upper Jurassic Morrison Formation of Colorado, Utah, and Wyoming; the Lower Cretaceous beds of the Kaiparowits Plateau of southern Utah; the Lance Creek beds near Hat Creek in eastern Wyoming; or the Hell Creek beds near Jordan, Montana. But finding and properly collecting more complete dinosaur remains should be undertaken by professionals and specialists who know how to get them out of the ground in a plaster jacket without destroying them, how to transport and prepare them, and who know how to place them in a proper scientific institution for study. A number of "summer paleontology camps" that advertise online can provide you with this experience without the expense and difficulty of transporting and storing dinosaur fossils.

CLASS AVES: BIRDS

Birds are among the most popular of all wildlife to study, and many people own birds as pets or for a hobby. Today there are over 10,000 known species of birds. Despite their great diversity, their fossil record is relatively poor because birds have delicate hollow bones that are easily broken. Forty-seven of the 155 living families of birds have no fossil record, and most of these 47 living families are known only from the Pleistocene. Most fossil birds are known from fragments of a few key bones. Complete articulated specimens have been found preserved at a only handful of extraordinary localities, such as the Cretaceous beds of Liaoning Province in China or the Eocene Messel lake beds of Germany. Bird fossils are also commonly found in extraordinary deposits such as La Brea tar pits, although the fossils are all jumbled and disarticulated and no two bones can be reliably associated. Bird fossils are not easily collected or owned, and most people have to be content with viewing specimens on display in museums.

One of the first bird fossils ever found was the extraordinary discovery of *Archaeopteryx* in the Upper Jurassic Solnhofen Limestone of Germany in 1861 (fig. 19.19A), just two years after Darwin predicted such discoveries

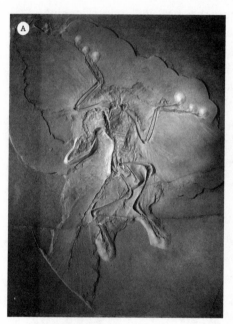

Figure 19.19 ▲

Some remarkable fossil birds. (*A*) The best of the 12 known specimens of *Archaeopteryx* from the Upper Jurassic Solnhofen Limestone, now on display in the Museum für Naturkunde in Berlin. (*B*) The gigantic flightless moas from New Zealand died out just a few hundred years ago when the Maoris hunted them to extinction. This is Sir Richard Owen, who first described the biggest of the moas, *Dinornis*. (*C*) The half-ton "elephant birds" of Madagascar also vanished when humans hunted them to extinction. (*D*) In the Eocene of North America (*Diatryma*) and Europe (*Gastornis*), enormous predatory birds hunted the much smaller mammals. (*E*) The absence of large mammal predators in South America during the Cenozoic allowed giant predator birds called phorusrhacids to evolve. (*F*) The enormous *Argentavis* compared to the largest living flying bird, the Andean condor. ([*A–C, E*] Courtesy of Wikimedia Commons; [*D*] photograph by the author; [*F*] courtesy of N. Tamura)

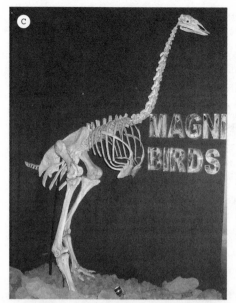

Figure 19.19 ▲
(*continued*)

in his book *On the Origin of Species*. *Archaeopteryx* is an ideal transitional fossil showing how birds evolved from their dinosaurian ancestors, especially dromaeosaur theropod dinosaurs such as *Velociraptor* and *Deinonychus*. Even though it had feathers, in almost every other respect *Archaeopteryx* had a primitive dinosaurian skeleton. It had a dinosaurian skull with peg-like teeth in the jaw (all living birds have a toothless beak), long bony fingers capable of grasping (the hand of living birds is fused into a single bone that supports the shafts of the wing feathers), a long bony tail (living birds have a short bony tail and feather shafts support their tail feathers), wrist bones identical to those found in *Velociraptor*, and the characteristic ankle joint found only in dinosaurs and pterosaurs. In fact, one specimen with faint feather impressions was mistaken for the small dinosaur called *Compsognathus*.

Since the discovery of *Archaeopteryx*, many more Mesozoic bird fossils have been found, especially in the Cretaceous deposits of China where they are complete with feather impressions and even their original color. These fossils show the various steps in evolution from birds like dinosaurs to birds with modern features. One group, the Enantiornithes ("opposite birds" in Greek), occupied most of the smaller bird niches in the Cretaceous, but they were extinct by the end of the Mesozoic. Another group, the ornithurines, were related to living birds, but they were overshadowed by the enantiornithines. These included a number of larger mostly marine birds of the Cretaceous, such as the shorebird *Gansus* from China, plus toothed birds from the western Kansas chalk beds, such the loon-like *Hesperornis* and the tern-like *Ichthyornis*. Most of these Mesozoic lineages vanished before the Cretaceous extinctions, leaving only the living branches of birds (Neornithes) to radiate and evolve rapidly in the early Cenozoic.

One of the most primitive living branches of birds are the flightless ratite birds, such as the ostrich of Africa, the rhea of South America, the emu and cassowary of Australia, and the kiwi of New Zealand. In addition, even larger ratites existed, such as the huge flightless moas of New Zealand that were up to 12 feet (3.7 meters) tall; they vanished only 400 years ago when Maoris hunted them to extinction (fig. 19.19B). Madagascar had the largest bird ever found, the "elephant bird" *Aepyornis* (fig. 19.19C), which weighed about 1,000 pounds (450 kilograms) and laid an egg almost a foot long that had a two-gallon capacity.

The rest of the modern birds (neognath birds) include the bulk of the 10,000 species and 54 families alive today. Among these lineages, there have been some surprising developments. After the large predatory dinosaurs vanished during the great Cretaceous extinction event, there were no large mammalian predators until the middle Eocene, about 40 million years ago. Into this void gigantic predatory birds evolved to reestablish the dinosaurian dominance over mammals. During the early Eocene, North America had *Diatryma* and Europe had *Gastornis*, both huge 7-foot-tall (2 meters) carnivorous flightless birds with sharp robust beaks for ripping apart smaller mammalian prey (fig. 19.19D). Likewise, for most of its Cenozoic history, South America had no large predatory mammals other than wolf-sized marsupials. Instead, the dinosaurs ruled the world again with huge 7- to 9.8-foot-tall (2 to 3 meters) predatory birds known as phorusrhacids that had long robust legs and sharp hooked beaks (fig. 19.19E).

Flying birds cannot grow as large as flightless ground birds because they must keep their bones and bodies light enough to fly. Nevertheless, there were some amazing examples of huge flying birds. The largest yet known was the condor-like *Argentavis* from the Miocene of Argentina (fig. 19.19F). It had a wingspan of 23 feet (7 meters) and weighed about 150 pounds (72 kilograms). It probably soared on the thermal updrafts along the Andes looking for food and carrion just as modern eagles and vultures and condors do. A huge seabird, *Pelagornis*, was found in the Oligocene beds excavated during the construction of the airport in Charleston, South Carolina. It had a similar wingspan, possibly as long as 24 feet (7.4 meters), but the long narrow wings of an albatross, and it probably only weighed about 88 pounds (40 kilograms).

CLASS MAMMALIA: MAMMALS

With about 5,400 living species, mammals are not as diverse as birds (10,000 living species) or bony fish (30,000 species) or insects (millions of species). People sometimes call the last 65 million years the Age of Mammals, but mammals are not the most numerous creatures during that time. Nevertheless, mammals have been the dominant medium and large animals on Earth ever since the extinction of the nonbird dinosaurs 65 million years ago. They include huge beasts like the elephant and the blue whale (the largest animal that ever lived), as well extinct gigantic mammoths and immense hornless rhinoceroses, and many extinct whales. Mammals have also taken over nearly all the important roles on land, from large herbivores to nearly all the predatory niches, to arboreal and digging roles, to ant-eating habitats. Independently of birds and insects, some mammals evolved flight (bats). Mammals (whales, seals, and sea lions) also became the dominant predators in the marine realm. Not only did mammals dominate the large animal niches, but they got small too. Although not as small as insects, the most diverse mammals are the tiny ones, especially rodents, rabbits, and insectivores. Some mammals are truly tiny, especially those that prey on insects. The living Etruscan shrew weighs only 0.07 ounces (2 grams) and is barely over an inch (2.5 centimeters) long. The smallest mammal known was the extinct *Batonoides vanhouteni* from the early Eocene (53 million years ago) of Wyoming. It weighed only 0.05 ounces (1.3 grams) and was about the size of an eraser on the tip of a pencil.

Living mammals are defined by a large number of features. For example, all living mammals have some kind of hair or fur on their bodies for insulation, even though some (like elephants and rhinoceroses) have lost most of it. All living mammals (except the platypus and echidna) no longer lay eggs but give live birth to their young, and all mammal females have mammary glands to nurse their young once they are born. All mammals are "warm blooded," or endothermic, using the energy of their food to keep their body temperature warm and relatively constant. They have a four-chambered heart, a diaphragm for pumping air in and out of the lungs, and the most sophisticated and largest brains of the entire animal kingdom, which is equipped with an expanded neocortex. Some mammals (such as humans, apes, elephants, and dolphins) have very complex behaviors as well.

Unfortunately, most of these characteristics are seen as behaviors or occur in soft tissue that does not fossilize well. To trace the origin of mammals in the fossil record, we look for a series of features in the skeletons of the fossils that show the transition from primitive reptile-like creatures (such the middle ear bones, the jaw hinge, the secondary palate in the mouth, specialized teeth, and many others) to true mammals.

These ancestors have long been called "mammal-like reptiles," but that term is inappropriate because these creatures have nothing to do with the lineage that leads to reptiles. Instead, they should be called

Figure 19.20 ▲

Reconstructions of some protomammals or synapsids (formerly but incorrectly called "mammal-like reptiles"). On the right in the background is the finbacked predatory "pelyco-saur" *Dimetrodon*, and on the left is the finbacked herbivorous "pelycosaur" *Edaphosaurus*. In front on the left is the huge predatory "therapsid" gorgonopsian *Gorgonops*, and behind it is the herbivorous "therapsid" dinocephalian *Moschops*. Behind the human is the dinocephalian *Estemnosuchus*, with the bizarre crests and tusks on its face. In the right front are the wolf-sized predatory cynodont *Cynognathus* and the cow-sized herbivorous dicynodont "therapsid" *Kannemeyeria*, with a toothless beak except for canine tusks. (Illustration by Mary Persis Williams)

"protomammals" or their proper name, Synapsida (fig. 19.20). The earliest synapsids (*Protoclepsydrops* and *Archaeothyris*) and true reptiles (*Westlothiana*) originated side by side in the Early Carboniferous as separate lineages, so at no time were synapsids ever "reptiles."

After the small Carboniferous lizard-like synapsids such as *Protoclepsydrops* and *Archaeothyris*, the first great radiation of protomammals occurred in the Early Permian with the familiar "fin-backed" creatures like *Dimetrodon* and *Edaphosaurus*, originally placed in the wastebasket group "pelycosaurs" (fig. 19.21A–B). These fossils are common in the Lower Permian beds around Seymour, Texas, and are still collected today. The tiger-sized *Dimetrodon* was the largest land predator the world had seen up to that time. Pelycosaurs vanished in the Late Permian, replaced by an even bigger evolutionary diversification of more advanced synapsids. These included large herbivores like the pig-sized dinocephalians, with thick skulls and weird bumps on their head, and the beaked dicynodonts, as well as bear-sized predators with huge fangs like the fearsome

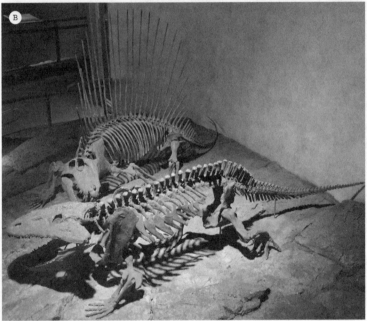

Figure 19.21 ▲

(A) The earliest synapsids or "pelycosaurs" included the finbacked *Dimetrodon*, the dominant predator of the Lower Permian red beds of northern Texas. (B) Fossils of *Dimetrodon*, attacking the contemporary temnospondyl *Eryops*. (C) Reconstruction of the gorgonopsians, huge predators of the Late Permian. (D) Skeleton of the huge Late Permian gorgonopsian *Gorgonops*. (Photographs by the author)

Figure 19.21 ▲
(*continued*)

gorgonopsians (fig. 19.21C–D). Most Late Permian protomammals vanished in the great Permian extinction, but several lineages survived, and during the Triassic they evolved into even more mammal-like creatures that had most of the mammalian features of the skeleton (fig. 19.22). Some, such as *Thrinaxodon*, have bony features and may have had hair and a diaphragm. By the Late Triassic, these advanced protomammals were very similar to true mammals, and it is difficult to decide where the protomammals end and true mammals begin.

There are several shrew-sized Late Triassic fossils that almost all paleontologists regard as true mammals because they had the mammalian jaw joint and middle ear bones, and their skeletons had all the other advances seen in mammals. From the Triassic through the rest of the Mesozoic (over 130 million years, twice as long as the Cenozoic Age of Mammals), mammals evolved and diversified rapidly in a world dominated by dinosaurs (which also originated in the Late Triassic). They remained small (mostly rat-sized or smaller) and probably hid in the undergrowth and came out mostly at night to escape being eaten by their dinosaurian overlords (fig. 19.23). If not for the great Cretaceous extinction event that wiped out the nonbird dinosaurs, mammals would still be tiny and hiding from dinosaurs, and we would not be here.

At the dawn of the Cenozoic Age of Mammals, mammals inherited a planet with no larger creatures ruling over them. During the next 15 million years, they underwent a huge evolutionary radiation from a handful of shrew-sized creatures to many different lineages. By the middle Eocene (about 47 million years ago), not only were some of them almost the size of elephants, but early bats were also flying in the skies, primitive whales lived in the oceans, and there were large predators the size of wolves, a wide spectrum of primates in the trees, as well as primitive anteaters and many kinds of burrowing mammals (fig. 19.24A–G).

This great evolutionary radiation of Cenozoic mammals produced three main groups that are still alive today. One branch, the monotremes, includes the living platypus plus the echidnas, or "spiny anteaters," of Australia and New Guinea. These are the only mammals that still lay eggs. The females do not have nipples but they do have mammary glands to nurse their young when they hatch, so the babies must lap up the milk from the mother's fur.

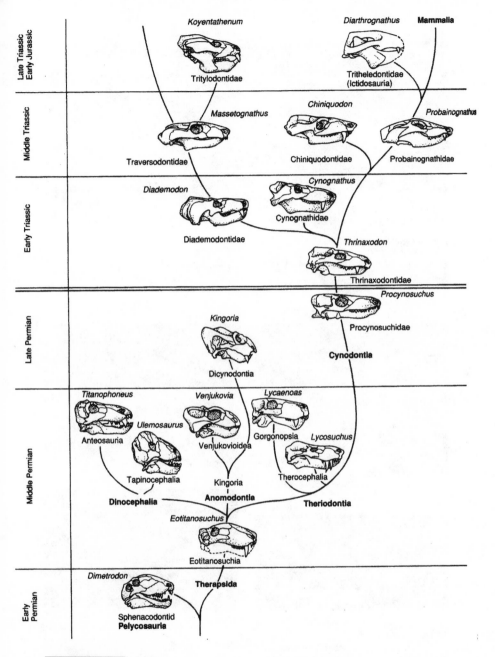

Late Triassic / Early Jurassic

Koyentathenum

Tritylodontidae

Diarthrognathus   **Mammalia**

Tritheledontidae
(Ictidosauria)

Middle Triassic

Massetognathus

Traversodontidae

Chiniquodon

Chiniquodontidae

Probainognathus

Probainognathidae

Early Triassic

Diademodon

Diademodontidae

Cynognathus

Cynognathidae

Thrinaxodon

Thrinaxodontidae

Late Permian

Kingoria

Dicynodontia

Procynosuchus

Procynosuchidae

Cynodontia

Middle Permian

Titanophoneus

Anteosauria

Ulemosaurus

Tapinocephalia

Dinocephalia

Venjukovia

Venjukovioidea

Kingoria

Anomodontia

Lycaenoas

Gorgonopsia   *Lycosuchus*

Therocephalia

Theriodontia

Eotitanosuchus

Eotitanosuchia

Early Permian

Dimetrodon

Therapsida

Sphenacodontid
Pelycosauria

Figure 19.22 ▲

Evolution of the synapsids, from the primitive "pelycosaurs" of the Early Permian, to the
"therapsids" of the Late Permian, and finally to the advanced cynodonts and mammals of
the Early Triassic. (Redrawn from several sources by Mary Persis Williams)

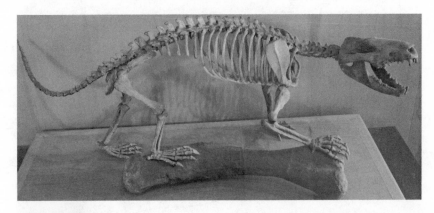

Skeleton of the larger Mesozoic mammal *Gobiconodon*, about the size of cat. Most Mesozoic mammals were the size of a shrew or mouse. (Photograph by the author)

A spectrum of different mammalian groups that had evolved by the middle Eocene, about 45 million years ago. (*A*) The lemur-like primate *Notharctus*. (*B*; color figure 13) The elephant-sized horned and tusked uintatheres *Eobasileus*. (*C*) The transitional whale *Ambulocetus*, the "walking swimming whale," which still had four limbs with webbed feet. (*D*) The primitive rhino-like group known as brontotheres. (*E*) The earliest elephant relatives were pig-like forms called *Moeritherium*. (*F*) Two of the earliest bat fossils, *Icaronycteris* and *Onychonycteris*. (*G*) One of the earliest horses, *Protorohippus*. (Photographs by the author)

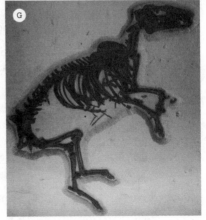

Figure 19.24 ▲

(*continued*)

Figure 19.25 ▲

A spectrum of extinct marsupials. In the foreground is the wolf-like sparassodont *Borhyaena*. In the extreme right foreground is the saber-toothed sparassodont *Thylacosmilus*. In the center foreground is the "marsupial lion" *Thylacoleo*. The giant short-faced kangaroo is *Procoptodon*. The huge creature behind the human is *Diprotodon*. The sloth-like creature with the proboscis in the left background is *Palorchestes*. (Illustration by Mary Persis Williams)

The second main branch of mammals are the marsupials, or the "pouched mammals," including opossums, kangaroos, wallabies, koalas, wombats, and many more species (fig. 19.25). Marsupials give live birth to their young, but the newborns are born prematurely as tiny bee-sized creatures with only a functional mouth and front limbs. Once they are born, they must crawl up the belly fur and crawl into their mother's pouch, where they clamp onto a nipple and finish their development until they can get around on their own.

Opossums and similar marsupials have long been successful in the Northern Hemisphere, but their greatest evolutionary success was in the southern continents of Australia and South America. Australia has had a long fossil history of marsupials, including some gigantic kangaroos twice as large as the living species and wombats the size of rhinos. Today nearly all the native mammals of Australia are familiar marsupials such as kangaroos, wallabies, koalas, wombats, and their kin, although many are being driven to extinction by competition from invading mammals (dingoes, rats, rabbits, sheep, goats, cattle) brought by humans and by human destruction of their habitat. Marsupials also dominated South America during most of the Cenozoic when it was an "island continent" isolated from mammals evolving in the rest of the world. Nearly all the large predators of South America were marsupials, including some that looked much like wolves or hyenas, and one pouched predator that was a good mimic of a saber-toothed cat (see fig. 19.25).

But the most diverse groups of mammals since the early Cenozoic are the placental mammals. Unlike marsupials, placental females carry their young to term inside their bodies, so they can be born almost fully functional. Some, like baby zebras or antelopes, must be able to stand and run with the herd within minutes of being born, or predators will get them. Others, like humans, are born relatively defenseless, but at least all of our organs are fully developed and functional even if the baby is not able to do much on its own at first.

I cannot review all 20-plus orders of fossil mammals, which are composed of many hundreds of genera (living and extinct) and thousands of species. Instead, I will focus on a few groups that the collector is likely to encounter in the field. Unlike other vertebrates, for which we must collect most of the skeleton in order to identify and understand them, fossil mammals can usually be recognized and identified by their teeth alone. Teeth are the most durable part of the vertebrate skeleton and have a dense covering of hard enamel that allows the tooth to survive being bashed around in the currents of rivers or the ocean. Unlike the simple peg-like teeth of most reptiles or amphibians (or the toothless birds), mammal teeth are highly specialized. The shape and crown pattern and cusps of the molars and premolars of the cheek teeth, in particular, are highly diagnostic not only of what animal the teeth came from but even what kind of food the animal ate, how old the individual was at the time of death, and what the environmental conditions were like when the mammal lived. A high percentage of mammal species are known only from their teeth and jaws and no other part of the skeleton. I will introduce a few of the better known or commonly collected North American fossil mammals and explain how they are fossilized and identified, including the features of their tooth crowns.

XENARTHRANS: SLOTHS, ARMADILLOS, AND ANTEATERS

The xenarthrans (formerly called the "edentates" or "toothless ones") are familiar to us from the living sloths, anteaters, and armadillos (fig. 19.26). They are exceptions to the rule about recognizing mammals by their teeth; sloths

Figure. 19.26 ▲

Examples of some of the extinct giant xenarthrans. The gigantic sloth (*middle*) is *Megatherium*; the sloth (*right front*) is *Megalonyx*, and the sloth (*left front*) is *Mylodon*. The armored creature (*middle left*) with the spiky tail is *Doedicurus*, and the giant armadillo relatives in the back are *Glyptodon* (*right*) and *Holmesina* (*left*). (Illustration by Mary Persis Williams)

and armadillos have simple peg-like teeth with no enamel coating, and anteaters are completely toothless. Xenarthrans are the most primitive group of living placental mammals, branching off during the Mesozoic before most of the other groups. They became isolated in South America and underwent most of their evolution there, before they began to cross Central America to North America in the late Miocene. At that point, huge ground sloths invaded North America, and they left their bones in many places. The biggest were elephant-sized ground sloths in South America that were 20 feet (6 meters) tall and weighed 3 tonnes. North America also had huge sloths that were only slightly smaller. The two living species of tree sloth found in Central and South America are remnants of a huge ground sloth radiation that have shrunk down in size. Each evolved from a different family of ground sloths, so the two-toed sloth is not that closely related to the three-toed sloth.

The other major group of xenarthrans are the armadillos (which are still common in the desert Southwest of the United States) and their extinct relatives, the glyptodonts. Glyptodonts and the very similar pampatheres looked like armadillos but were the size of a Smart car, with a domed shell over 6 feet (2 meters) long, and weighing up to 2 tonnes. Some had a spiked tail club, and others had only armor on their tail and over their head.

Although complete glyptodonts are rare, the individual pieces of their body armor are easily fossilized and quite often found in Pliocene and Pleistocene fossil beds all over the Americas.

PROBOSCIDEA: MASTODONTS AND MAMMOTHS

The mastodonts and their ancestors originated in the Paleocene of Africa; they were pig-sized beasts that had only the beginnings of tusks and possibly a short proboscis (fig. 19.27; see also fig. 19.24E). In the early Miocene (about 19 million years ago), they escaped Africa, quickly spreading across Eurasia, and finally across the Bering land bridge to North America. Most important Miocene and Pliocene mammal localities will yield mastodont fossils or fragmentary bone scraps because their bones are large and durable. The majority of the Miocene proboscideans are known as gomphotheres; they had long narrow jaws and straight upper and lower tusks protruding from their jaws. There were also shovel-tusked mastodonts, whose lower tusks were merged into a broad scoop or shovel-like shape. By the Pliocene, more specialized proboscideans appeared in Eurasia and Africa, and some migrated to the Americas. The most familiar of these are the mammoths, which looked much like modern elephants except they tended to be bigger and their tusks had a long inward curve. Numerous species of mammoth roamed North America during the Pleistocene, including the smaller woolly mammoth in the polar regions, and the much larger (but not hairy) imperial mammoth, found in most southern Pleistocene localities, such as at La Brea tar pits.

Proboscideans are easily recognized from their teeth alone. Like their close relatives the manatees and other sea cows, proboscideans have what is known as horizontal tooth replacement. Instead of the adult teeth pushing out the baby teeth from below (as in most mammals), the new teeth erupted from the back of the jaw and pushed the old worn teeth off the front (fig. 19.28A). Mastodonts have very distinctive large molars with multiple rounded conical cusps, which occasionally are joined together to form cross-crests (fig. 19.28B). Elephants and mammoths, on the other hand, had huge molars made of tightly folded ridges of enamel and dentin, which wore down to form a ridged grinding surface. Each molar was so large, and the jaw was so short, that an elephant or mammoth had only one tooth (or maybe two) on each side of its upper and lower jaws at any given time. These teeth are easily recognized and identifiable, even when they are broken and no longer attached to the jaw.

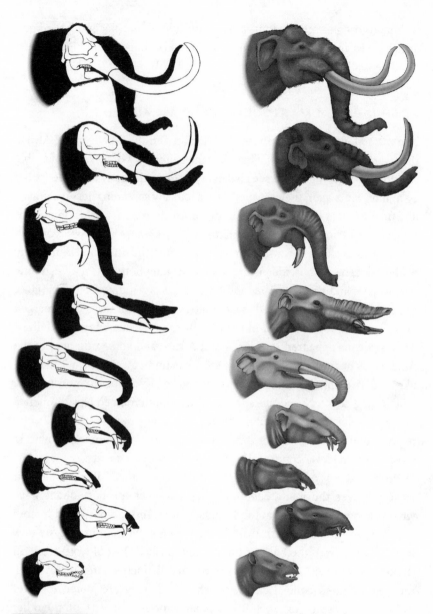

The evolution of proboscidean skulls from the Eocene to the Pleistocene. At the bottom is
Phosphatherium, followed (*bottom to top*) by *Numidotheriun, Moeritheriun, Palaeomast-
odon, Phiomia, Gomphotherium, Deinotherium, Mammut* (American mastodon), and at the
top, *Mammuthus*, the mammoth. (Illustration by Mary Persis Williams)

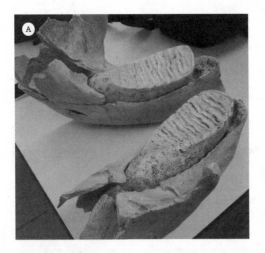

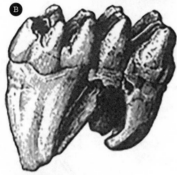

Figure 19.28 ▲

All proboscideans have horizontal tooth replacement. One enormous molar on each side of the jaw (and on the upper jaw as well) is in use at a time, and it is pushed forward and breaks off the front as it wears down, only to be replaced by another tooth behind it. (A) Top view of a mammoth jaw, showing one molar tooth in occlusion, and another unworn tooth behind it in the crypt in the back of the jaw. (B) The differences between a mastodon molar (left) with the low rounded cusps, and a mammoth or elephant molar, a solid grinding tooth made of thick bands of enamel separated by softer dentin, creating a grinding surface. ([A] Photograph by author; [B] courtesy of Wikimedia Commons)

CARNIVORES: DOGS, CATS, BEARS, AND THEIR RELATIVES

Preying upon other mammals is the order Carnivora, the flesh-eating mammals (fig. 19.29). Today they include dogs, cats, bears, weasels, skunks, raccoons, hyenas, civets, and mongoose, as well as seals, sea lions, and walruses. Most meat-eating mammals have distinctive jaws

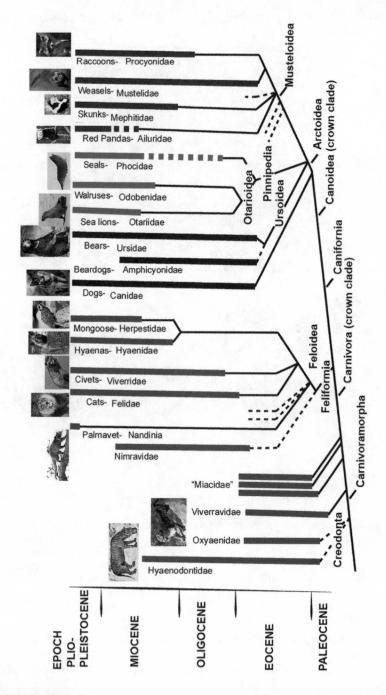

Figure 19.29 ▲

Family tree of the creodonts and carnivorans. (Redrawn from several sources by Mary Persis Williams)

and teeth, with long sharp canines, for stabbing and tearing into their prey, and cheek teeth that are shaped like slicing blades, for cutting the meat up before they swallow it.

The earliest predators were an extinct group known as creodonts (fig. 19.30A–B), which had a much more primitive combination of teeth, small brains, robust limbs, and were mostly wolf-sized to lion-sized ambush predators. Creodonts were the dominant predators in Paleocene and Eocene beds, but by the middle and late Eocene, members of the living order Carnivora had appeared and began to crowd them out. The earliest dogs from the late Eocene (*Hesperocyon*) looked more like weasels, and the dog family went through an extensive evolutionary history in North America, with one branch developing hyena-like crushing jaws and teeth for breaking bones and scavenging. Living alongside them in the late Eocene and Oligocene were cat-like creatures known as nimravids, or "false cats" (fig. 19.30C). Although their skulls and teeth are extraordinarily cat-like

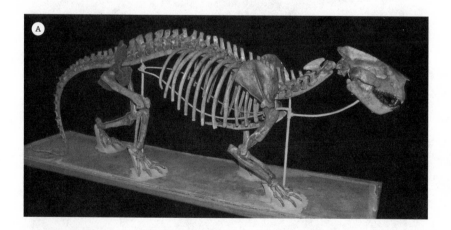

Figure 19.30 ▲

Some examples of extinct carnivorans. (A) The jaguar-sized creodont *Patriofelis*, one of the earliest large mammalian predators. (B) One of the last of creodonts, the wolf-like *Hyaenodon*, common from the Eocene through the Miocene around the world. (C) The "false saber-tooth cat" or nimravid *Hoplophoneus* from the Eocene and Oligocene. (D) The true saber-tooth cat, *Smilodon*. (E) The bone-crushing borophagine dog, *Epicyon*. (F) The enormous bear dog *Amphicyon*. (G; color figure 14) The short-faced bear, one of the largest land predators that ever lived; it is bigger than any living bear. ([A–B] Courtesy of Wikimedia Commons; [C–G] photographs by the author)

Figure 19.30 ▲
(*continued*)

Figure 19.30 ▲
(*continued*)

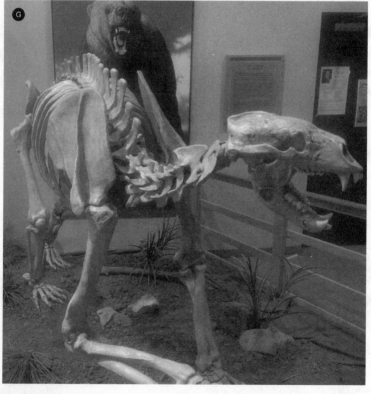

Figure 19.30 ▲
(continued)

(like *Dinictis*), with some even developing saber-like teeth (*Hoplophoneus*, *Nimravus*, and *Eusmilus*), the details of their skull region show that they were not true cats at all, and they even may be more closely related to dogs. Whatever their relationships, they represent a truly extraordinary example of evolutionary convergence. When the nimravids vanished from North America around 26 million years ago, there were no cat-like creatures to replace them. This resulted in a "cat gap," a total absence of cat-like predators in North America for 7.5 million years in the late Oligocene and early Miocene, until true cats arrived from Eurasia about 18.5 million years ago (fig. 19.30D).

By the early Miocene, a variety of primitive weasel, bear, and raccoon relatives had evolved, often migrating back and forth between North America and Eurasia across the Bering land bridge. The dog family, on the other hand, remained largely confined to North America until later in their evolution. They spread to South America about 4 million years ago to become the ancestors of the South American dog radiation (including bush dogs and maned wolves), and they spread to Eurasia as well. In North America, one dog family, the borophagines, evolved extremely powerful skulls with bone-crushing teeth and performed the role of predator-scavengers that hyenas occupied in the Old World (fig. 19.30E).

Another important group were the extinct amphicyonids, nicknamed the "bear dogs." They are distantly related to both bears and dogs but are not members of either living group. Some, like *Amphicyon* (fig. 19.30F), were gigantic wolf-like predators larger than the largest bear. By the late Miocene, all the archaic predators such as bear dogs and creodonts were completely extinct. Most of the predatory ecological niches were filled by cats, dogs, bears, and their relatives. During the Ice Ages, North America was home to some extraordinary predators, including saber-toothed cats, giant short-faced bears bigger than the largest living bears (fig. 19.30G), huge lion-like cats bigger than any lion, and many different kinds of dogs, including the fearsome dire wolves, which are the most common fossil at places like La Brea tar pits.

ORDER PERISSODACTYLA: ODD-TOED HOOFED MAMMALS

Today, only five species of tapirs, five species of rhinos, and seven species of wild horses, asses, and zebras (fig. 19.31) survive on Earth, but through much of geologic history these were the dominant large hoofed mammals

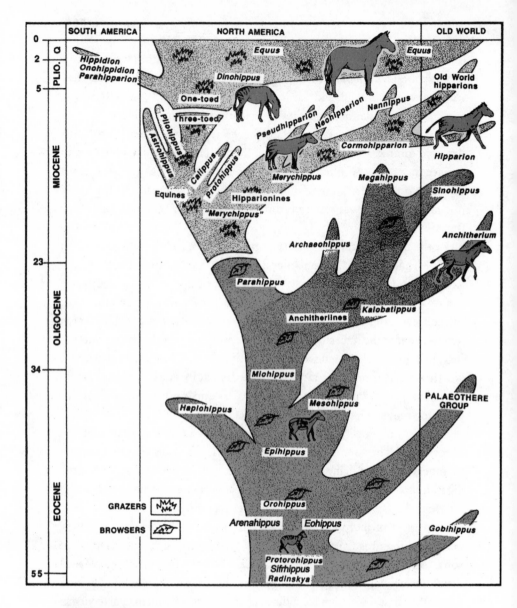

Figure 19.31 ▲

The evolution of horses. Horses began as collie-sized creatures with four fingers on their hands and three toes on their feet (*Protorohippus*), but by the Oligocene they had become larger and used only three toes. All of these horses ate leaves and other browse (*lower branches with leaves in background*), as did their descendants, the anchitherine horses (*fourth branch from bottom to the right*), which reached the size of modern horses but still had low-crowned teeth for eating leaves. The main lineage of horses (*upper branches with grasses in background*), however, developed high-crowned teeth for grinding tough, gritty grasses, which wears teeth down quickly. There were several branches of three-toed horses in the Miocene, including the merychippines, the hipparionines, the calippines, and others. Like the anchitheres, some of these migrated to the Old World in the middle and late Miocene. All were extinct by the end of the Miocene, leaving only the one-toed horse lineage that was ancestral to modern horses (genus *Equus*). (Drawing by C. R. Prothero)

not only in Africa and Eurasia but especially in North America (where none of them are native today). They are members of the order Perissodactyla, or the odd-toed hoofed mammals. They get this name because their hands and feet have only one (horses) or three (rhinos and tapirs) fingers and toes, and the axis of symmetry runs through the middle finger and toe.

Perissodactyls originated from Asian relatives like *Radinskya* of the Paleocene of China and Mongolia, but in the early Eocene they migrated between all the northern continents. If you go collecting in the lower Eocene beds of the Bighorn Basin of Wyoming or the San Juan Basin of New Mexico, the most common larger mammal fossils you will find are the teeth of early horses, tapirs, and rhinos. About 55 million years ago, they were so similar to one another that even experts have a hard time telling them apart, both in their skull and skeletons, and even in their teeth. Primitive horse teeth from *Protorohippus* (formerly *Eohippus* or *Hyracotherium*) are particularly common (fig. 19.32A), and they are extremely similar to the earliest tapiroid, *Homogalax*.

Horses continued to be a largely North American group through the rest of the Cenozoic, except for a few lineages that managed to escape to Eurasia in the Miocene. By the late Eocene to early Oligocene, they had developed into collie-sized three-toed horses such as *Mesohippus* and *Miohippus* (fig. 19.32B), which are among the most common fossils in the Big Badlands of South Dakota and in similar beds in Nebraska, Wyoming, and Colorado. During the Miocene, horses underwent a huge diversification in North America, with many different lineages living side by side. The anchitherine horses retained primitive low-crowned teeth and three toes, so they must have continued to browse on soft leaves, even though some of them were as big as a modern horse. But most horses in the Miocene became specialized grazers, or grass eaters, as the savannah grasslands expanded. Not only did they develop long slender limbs with only tiny side toes to enable them to run across the savannah, but they also evolved teeth with very tall crowns that could keep on growing out even as they were worn down by the grit in their grassy diet. Isolated high-crowned horse teeth of Miocene horses like *Merychippus* and hipparions are among the most common and age-diagnostic fossils in those beds (fig. 19.32C). Most of these lineages of horses vanished at the end of the Miocene, and only the one-toed horse *Dinohippus* and the modern genus *Equus* are found in the Pliocene and Pleistocene. Horse teeth are among the most common mammal fossils found in Ice Age beds all over North America.

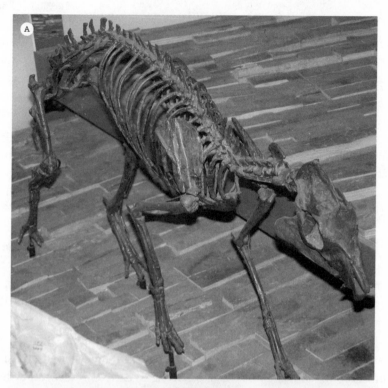

Figure 19.32 ▲

Some representative extinct perissodactyls. (*A*) The early Eocene horse *Protorohippus* and (*B*) the Oligocene horse *Miohippus*. (*C*) Horses evolved from being small and three-toed to large with a single central toe. Tapirs are easily recognized because the bones around their nasal opening have retracted, allowing the muscles for the proboscis to attach. (*D*) This middle Eocene early tapir *Hesperaletes* was the first to show this nasal bone retraction. (*E*) The Ice Age Asian giant tapir *Megatapirus*. (*F*) The running rhinocerotoid *Hyracodon*. (*G*) The first rhinos to bear paired horns on their noses, known as *Menoceras*. (*H*) The hippo-like Miocene rhinoceros *Teleoceras*. (Photographs by the author)

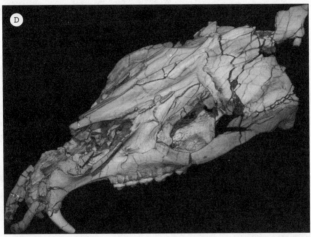

Figure 19.32 ▲
(continued)

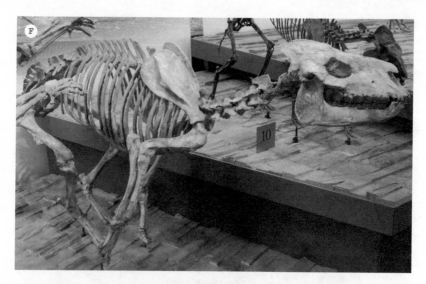

Figure 19.32 ▲
(continued)

Tapiroids were relatively common in the lower Eocene beds of North America, but they are rare fossils after that time. Nonetheless, they show a remarkable evolutionary history of different forms, many of which had developed the long proboscis, or snout, as early as the late middle Eocene (fig. 19.32D–E).

Rhinoceroses had an even better fossil record throughout most of the Cenozoic of North America. Most fossil rhinoceroses were hornless, and only a few developed horns, especially the African and Asian forms. Unlike other mammals, rhinoceros horns are not made of bone but of dense hair fibers glued together. They rarely fossilize, but they leave distinctive roughened scars on the snout and skull where they were attached. In the absence of the horns, we recognize most fossil rhinos by other features of their teeth and skeleton, especially their upper molars, which had a pattern of crests resembling the Greek letter "pi" (π). There were three main lineages of rhinoceros relatives: the hippo-like aquatic amynodonts (best known from *Amynodon* from the Uinta Basin of Utah and *Metamynodon* from the Big Badlands); the long-legged Great Dane–sized hyracodonts (*Hyracodon* is a common Big Badlands fossil; fig. 19.32F); and the true rhinoceroses, or Family Rhinocerotidae. The other two groups vanished by the end of the Oligocene in North America, but rhinocerotids had multiple waves of diversification after immigrants from Eurasia crossed over the Bering land bridge. In the Oligocene, the typical rhinos were *Subhyracodon* and *Diceratherium*. In the early Miocene, these were replaced by the cow-sized pair-horned rhino *Menoceras* from Eurasia (fig. 19.32G), the most common fossil in the Agate Springs fossil beds in Nebraska. By the late early Miocene, a new wave of rhinos came over from Eurasia: the primitive aceratherine rhinos such as *Aphelops* and *Peraceras*, which may have had a short proboscis and mostly browsed leaves; and the short-limbed, barrel-chested, hippo-like *Teleoceras*, which is extremely common in Miocene beds all over North America (fig. 19.32H). At the end of the Miocene, rhinoceroses vanished completely from North America, as did most of the groups that dominated the Miocene savannahs, probably because climate change in the Pliocene had made their habitats into a much colder and drier steppe grassland (fig. 19.33).

These three groups—horses, rhinos, and tapirs—are the only surviving perissodactyls. But during the Eocene, the largest and most spectacular

Figure 19.33 ▲

A stampede of rhinocerotoids: the small ancestral *Hyrachyus* (*right foreground*); the run-ning hyracodont rhino *Hyracodon* (*far left foreground*) and its relatives, which included the hyracodont *Juxia* from China (*middle rhino far right*) and the immense *Paraceratherium* from Asia, the largest land mammal known (*background*); and the hippo-like amynodont rhino known as *Metamynodon* (*left of human*). The remaining images are members of the true rhinos, Family Rhinocerotidae: the two-horned rhino *Menoceras* (*middle left*), the earli-est rhino to develop horns; the hairy rhino is the woolly rhino *Coelodonta* (*right foreground*); and the huge one-horned rhino is *Elasmotherium* (*center background*). The hippo-like rhino to the right of it is *Teleoceras*, and the small rhino charging the human is *Meninatherium*. (Illustration by Mary Persis Williams)

mammals of North America were the brontotheres or titanotheres (fig. 19.34). The early Eocene fossils *Lambdotherium* and *Palaeosyops* look much like the earliest horses, rhinos, and tapirs. By the middle Eocene, however, they were rhino-sized creatures with short, paired knob-like horns on their noses (see fig. 19.24D). In the upper Eocene beds of the Big Badlands, brontotheres reached the climax of their evolution. They were huge elephant-sized creatures with long curved blunt bony horns on their noses. They were by far the largest and most spectacular mammals on Earth up to that point. At one time, dozens of names, such as *Bronto-therium, Titanotherium, Brontops, Menodus, Allops, Titanops,* and *Menops,* were applied to these spectacular skulls, but now they are all placed in

Figure 19.34 ▶

Brontothere evolution portrayed as a simple linear process, advocated by Osborn in the early twentieth century. (After H. F. Osborn, 1929, *The Titanotheres of Ancient Wyoming, Dakota, and Nebraska*; U.S. Geological Survey, Washington, DC)

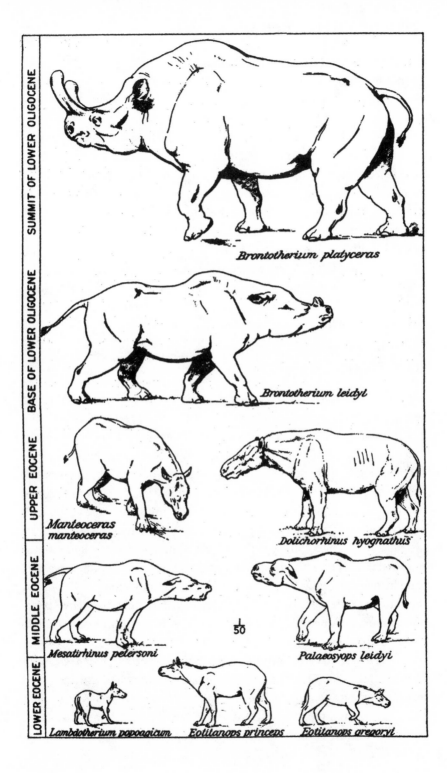

SUMMIT OF LOWER OLIGOCENE

BASE OF LOWER OLIGOCENE

UPPER EOCENE

MIDDLE EOCENE

LOWER EOCENE

Brontotherium platyceras

Brontotherium leidyi

*Manteoceras
manteoceras*

Dolichorhinus hyognathus

Mesatirhinus petersoni

Palaeosyops leidyi

Lambdotherium popoagicum

Eotitanops princeps

Eotitanops gregoryi

50

one genus, *Megacerops*, and the other names are no longer valid. Although brontotheres look vaguely like rhinoceroses, their horns are made of solid bone with rounded knobs (not matted hair, like rhino horns). At one time, scientists thought that brontotheres used their horns for head-to-head ramming, like bighorn sheep. But the bone of the nose and face supporting the horn is weak and spongy, and it could not have withstood huge impact stresses. It is more likely that the horns were used for display, and for head-to-head wrestling and sparring for dominance, as is true of most horned antelopes today. Brontotheres were the largest and most dominant animals on Earth by the end of the Eocene, but they vanished before the Oligocene began in both North America and Asia. They were the major victims of the great Eocene-Oligocene extinction event, when North American forests were replaced by dry scrublands. Brontotheres had primitive, low-crowned molars, suggesting that they could only browse on softer leaves, and those nearly vanished in the early Oligocene.

ORDER ARTIODACTYLA: EVEN-TOED HOOFED MAMMALS

The most common and diverse large mammals today are the artiodactyls, or the even-toed hoofed mammals (fig. 19.35). These include not only pigs and hippos, but also camels, deer, giraffes, pronghorns, cattle, sheep, goats, and antelopes, as well as many extinct groups. They get their name from the fact that their hands and feet have "cloven hooves." They have two or four fingers or toes, and the axis of their hands and feet runs between the middle (third) digit and the fourth digit (equivalent to the ring finger) instead of down the middle of the middle digit (as in perissodactyls). Primitive artiodactyls and pigs and hippos bear most of their weight on digits 3 and 4, have small side digits 2 and 5, and have no thumb or big toe. More advanced artiodactyls have only the two central digits, greatly elongated, to allow them to run faster. Artiodactyls also have an extremely distinctive set of ankle bones, and many other features of the skull, teeth, and skeleton help us recognize them.

The first artiodactyls from the early Eocene, such as *Diacodexis*, were about the size and shape of a large rabbit or a tiny antelope, with long hind legs for hopping and leaping. Even though they had a rabbit-like body, their skull, teeth, and limb bones clearly show they are primitive artiodactyls. North America did not have any native hippos or pigs, but the pig-like

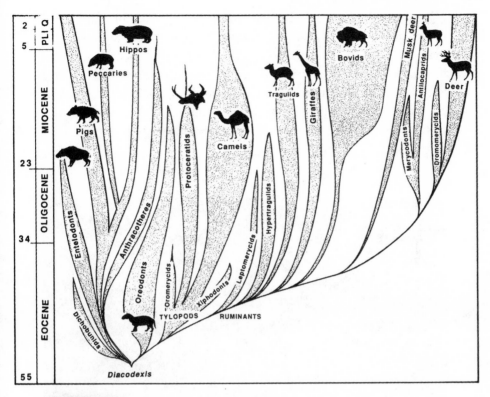

Figure 19.35 ▲

The evolutionary history of even-toed hoofed mammals, or artiodactyls. (Drawing by C. R. Prothero)

peccaries evolved here through most of their evolution until they spread to Latin America in the Miocene.

The most commonly fossilized artiodactyls in North America (and the most easily collected of all mammal fossils and widely available on the commercial market) are the oreodonts. This strictly North American group was about the size of a sheep or a pig and was distantly related to camels; however, it vanished during the late Miocene. During the Eocene and Oligocene and early Miocene, they were by far the most abundantly preserved of all larger mammals. The primitive ones, such as *Protoreodon*, occurred in huge numbers in the middle Eocene beds of San Diego or Utah. In the upper Eocene and Oligocene beds of the Big Badlands, the most commonly collected fossils are skulls of the oreodont *Merycoidodon* (fig. 19.36A), the dwarfed oreodont *Miniochoerus* (fig. 19.36B), or the larger late Oligocene *Megoreodon* (fig. 19.36C). They are easily recognized by many distinctive features of the skull and by their distinctive teeth with half-moon-shaped crescentic ridges. This selenodont tooth crown shape is typical of all advanced artiodactyls. During the Miocene, oreodonts diversified even further, and some became quite pig-like or even tapir-like, with a proboscis, such as *Brachycrus*, *Promerycochoerus*, and *Merycochoerus*

Figure 19.36 ▲

Oreodonts are among the most commonly collected fossils in the North American Eocene, Oligocene, and Miocene beds, and they are sold by many rock shops and fossil vendors around the world. (*A*) The common Big Badlands oreodont *Merycoidodon culbertsoni*. (*B*) Tray of the smaller but also very common Big Badlands oreodont *Miniochoerus*. (*C*) The late Oligocene oreodont *Megoreodon*. (*D*) The pig-like early Miocene oreodont *Promerycochoerus* had a proboscis. (*E*) Restorations of a diverse array of oreodonts: In the right foreground is the tiny form of *Miniochoerus*; behind it is the long-legged *Merychyus*, with high-crowned grazing teeth; and behind that is *Agriochoerus*, with the primitive long tail and claws instead of hooves. In the left foreground is the rabbit-sized leptaucheniid *Sespia*; behind it is the tapir-like *Brachycrus*; behind it is *Leptauchenia*, with the eyes and ears high on its head and extremely high-crowned teeth; In the left background is the common oreodont *Merycoidodon*. ([*A–D*] Photographs by the author; [*E*] illustration by Mary Persis Williams)

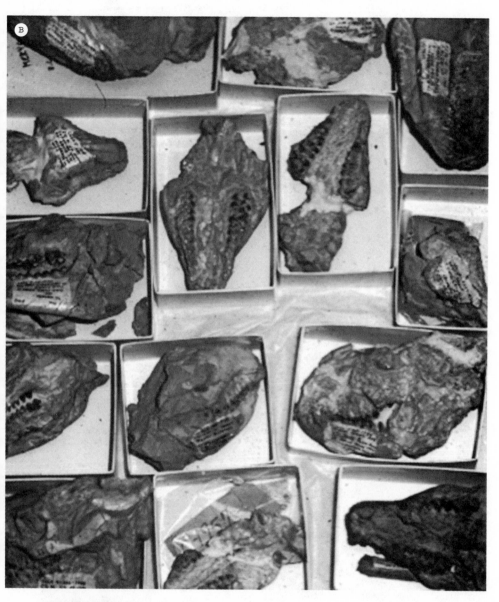

Figure 19.36 ▲
(*continued*)

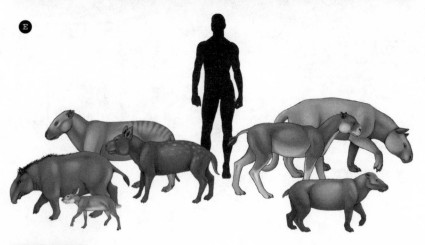

Figure 19.36 ▲
(*continued*)

(fig. 19.36D). Others had short slender limbs, large ears, and high-crowned teeth (*Leptauchenia*), and still others like *Merychyus* were delicate, fast runners (fig. 19.36E). They vanished completely in the late Miocene, leaving no descendants or even any close relatives.

Another extremely diverse and common American group is the camels (fig. 19.37A–B). Most fossil rhinos did not have horns, and most fossil camels did not have humps. In fact, four of the six living species of camels today (llamas, alpacas, vicuñas, and guanacos of South America) do not have humps. Only the African dromedary and the Asian Bactrian camel have humps, and they are specialized side branches of the camel family.

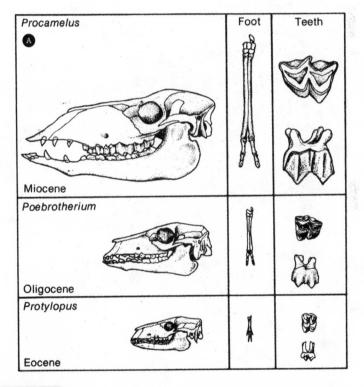

Figure 19.37 ▲

Camel evolution. (*A*) Simplified view of camel evolution from the early twentieth century, showing the increase in size of the skull and snout length, limb length and fusion, and higher crowned teeth. (*Protylopus* is not a camel; it is an oromerycid similar to the most primitive camels). (*B*) A modern view of camel evolution and phylogeny. ([*A*] Modified from several sources; [*B*] illustration by C. R. Prothero)

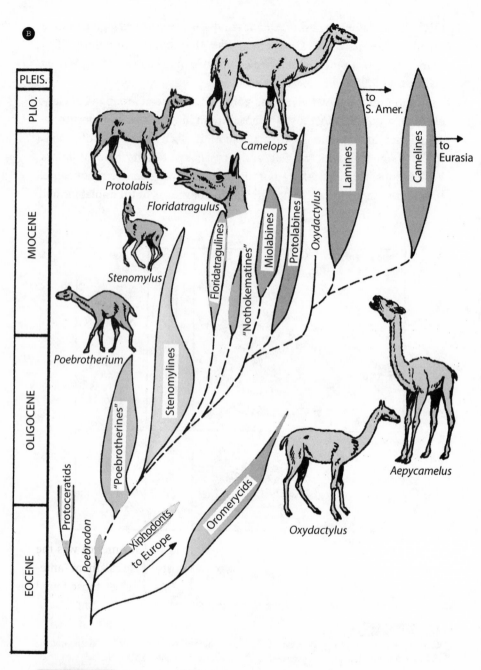

PLEIS.
PLIO.
MIOCENE
OLIGOCENE
EOCENE

Camelops

Protolabis

Floridatragulus

Stenomylus

Poebrotherium

Aepycamelus

Oxydactylus

to S. Amer.

to Eurasia

Lamines

Camelines

Floridatragulines

"Nothokematines"

Miolabines

Protolabines

Oxydactylus

Stenomylines

"Poebrotherines"

Protoceratids

Poebrodon

Xiphodonts to Europe

Oromerycids

Figure 19.37 ▲
(*continued*)

Camels first appeared in the late middle Eocene with *Poebrodon* in Utah and southern California, and by the late Eocene and Oligocene camels like *Poebrotherium* (fig. 19.38A) and *Paratylopus* were extremely common in the Big Badlands. In the Eocene, when most artiodactyls still had primitive low-crowned teeth, camels already had the highest-crowned teeth around, and they remained so throughout their history. During the Miocene, camels diversified into many groups, including the tiny gazelle-like stenomylines with their extremely high-crowned molars (fig. 19.38B), the long-necked "giraffe camels" (fig. 19.38B) that performed the roles of giraffes in the American savannah, and the shorter-limbed miolabine and protolabine camels (fig. 19.38C). Near the end of their evolution, camels became huge and heavy bodied, as their names *Titanotylopus* (fig. 19.38D) and *Gigantocamelus* suggest. Camels are the most common hoofed mammal fossils found in most of the Miocene beds of North America except for horses, and the high-crowned selenodont teeth and long limb bones of camels are very distinctive.

The more advanced artiodactyls are all ruminants, with a four-chambered stomach and the ability to chew their cud and survive on small amounts of high-quality vegetation. Ruminants first appeared in the late Eocene and Oligocene with the tiny hornless deer-like fossil *Leptomeryx*, a common fossil in Big Badlands collections. In the Miocene of North America, there was a great radiation of native blastomerycines, which are related to the living musk deer, as well as another radiation of deer-like fossils known as palaeomerycids or dromomerycids. Even though they look vaguely deer-like with many different horn combinations, they are completely extinct with no descendants, and their horns are more like giraffe horns than like the antlers of deer, which are shed by the bucks each year and regrown.

Another common Miocene ruminant group in North America was the pronghorns, or the antilocaprids. Even though most people call them "antelopes," they are not closely related to true antelopes such as those found in Africa and Asia. True antelopes are close relatives of cattle, sheep, and goats. Only one species of pronghorn, *Antilocapra americana*, still races across American grasslands, but during the Miocene there was a huge evolutionary radiation of pronghorns, some with remarkable varieties of horns on their heads. Pronghorn fossils, including their high-crowned teeth but especially their peculiar horns, are typical of the American Miocene in

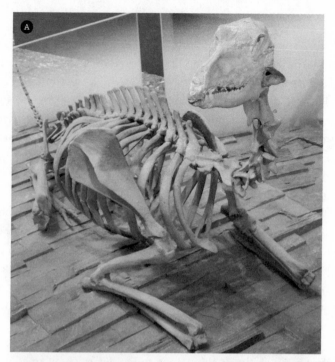

Figure 19.38 ▲
Some well-known fossil camels: (A) the late Eocene–early Oligocene Big Badlands camel *Poebrotherium*; (B) two late Oligocene-Miocene camels, the gazelle-like *Stenomylus* (*foreground*) and the "giraffe-camel" *Aepycamelus* (*background*); (C) the Ice Age lamine camel *Camelops* from La Brea tar pits; and (D) the giant camel *Titanotylopus*. (Photographs by the author)

Figure 19.38 ▲
(*continued*)

many places. Both the American musk deer and the dromomerycids vanished at the end of the late Miocene, and pronghorns nearly became extinct when the climate changed and the savannahs were replaced by the cold dry steppe habitats of the Pliocene.

Eventually, true deer migrated here from Eurasia, followed by some of the cattle relatives, such as *Bison* and musk ox, both of which were very common here in the Ice Ages. By the end of the Ice Ages, the dominant artiodactyls in North America were the llamas and the gigantic camels such as *Titanotylopus* and *Gigantocamelus*, as well as the bison. One of these, *Bison latifrons*, was a huge bison with giant horns that spanned over 7 feet (2.1 meters) from tip to tip (fig. 19.39). The great extinction at the end of

Figure 19.39 ▲

The giant Ice Age bison with a more than 7-foot horn span, *Bison latifrons*. (Photograph by the author)

the Ice Ages that wiped out most of the large mammals—from mastodonts and mammals to saber-toothed cats and ground sloths—also killed off horses, camels, and most of the other large Ice Age mammals. Only deer, bison, and pronghorns remained to roam our prairies and forests in the last 10,000 years.

PALEOBOTANY

Animal fossils are far more popular, but there are many places in the United States where abundant plant fossils can easily be collected as well. Like many types of animal fossils, only parts of the whole organism are found for most plant fossils: leaves, stems, trunks, roots, seeds, and so on. In many cases, paleobotanists (paleontologists who specialize in plant fossils) cannot conclusively show which leaf fossil belongs with which kind of petrified wood, but in rare cases, the entire plant from roots to leaves is preserved (fig. 20.1). In some places, especially ancient coal swamps and at the bottom of stagnant lakes, we find preserved plants in abundance. Other places, such as the Big Badlands of South Dakota, have preserved an amazing array of fossil animals—from tortoises and snails to many different kinds of mammals—but little evidence of the fossil plants that must have fed these creatures is found beyond the root casts and seeds of the hackberry bush (*Celtis*), which are calcified and were practically stone while the plant was alive.

In some cases, the preservation is extraordinary. The famous Miocene *Clarkia* beds of western Idaho are so poor in oxygen that when you split open a shale slab the 15-million-year-old leaves are still green. As they react to the oxygen in the air, they turn brown and can blow away in a matter of minutes if not quickly sealed in a preservative.

These early plants are important because they are the base of the food chain for almost all life, and they provide our only source of oxygen. Plant life made Earth habitable for other life forms, something that probably has not happened on any other planet. The earliest photosynthetic bacteria

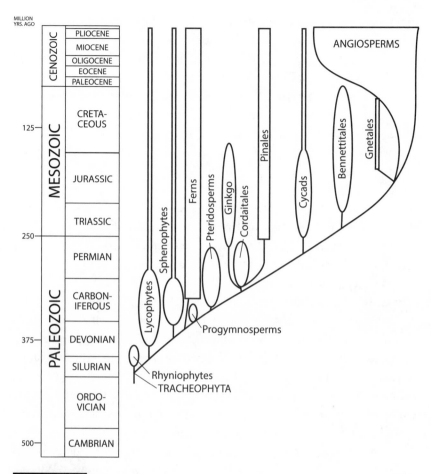

MILLION YRS. AGO

CENOZOIC	PLIOCENE
	MIOCENE
	OLIGOCENE
	EOCENE
	PALEOCENE

MESOZOIC — CRETA-CEOUS (125), JURASSIC, TRIASSIC (250)

PALEOZOIC — PERMIAN, CARBON-IFEROUS, DEVONIAN (375), SILURIAN, ORDO-VICIAN, CAMBRIAN (500)

ANGIOSPERMS

Lycophytes · Sphenophytes · Ferns · Pteridosperms · Ginkgo · Cordaitales · Pinales · Cycads · Bennettitales · Gnetales

Progymnosperms

Rhyniophytes
TRACHEOPHYTA

Figure 20.1 ▲

Evolution of land plants, showing the early diversification of spore-bearing plants, including horsetails (sphenophytes), ferns, and club mosses (lycopods). The first seed-bearing plants were gymnosperms, including seed ferns, cycads, and conifers. In the Early Cretaceous, there was a huge radiation of flowering plants, and they dominate most of the land plant diversity today. (Illustration by Mary Persis Williams)

spent almost 1.5 billion years pumping out oxygen after they first evolved about 3.8 billion years ago. It was not until about 2.4 to 1.8 billion years ago that enough oxygen had accumulated for the atmosphere to change from oxygen-starved to oxygen rich enough for animals to evolve. Without plants, Earth would have been as lifeless as Mars, which has no great biomass of plants to make oxygen and allow animals to grow. Likewise, those who imagine intelligent life on other planets forget how our planet only has life because of its perfect position between frozen Mars and scorching Venus, so it is not too hot and not too cold. Just finding a planet with the right temperature is not enough. It must have oceans of liquid water and some kind of organism that can produce oxygen before life like that found on Earth could arise.

For nearly all of life's history on Earth, until about 400 million years ago, the surface of the land was completely barren, exposed to the forces of weathering and erosion with no plants on the surface other than lichen-like organisms. Once large plants with tree-like shapes evolved in the Devonian, they transformed the land, making it habitable for the first animals to crawl out of the water. The millipedes were the first to do this, followed by scorpions, spiders, insects, and eventually the first amphibians.

Once forests with lots of woody biomass evolved in the Carboniferous Period, the atmosphere was transformed. Plants pumped out so much oxygen that gigantic insects and millipedes could evolve. The plants absorbed carbon dioxide, locking it into the earth's crust as coal. Apparently, there were no termites or other wood-digesting insects yet, so huge volumes of plant matter went directly to coal without the usual decomposers that break up decaying plant matter. Eventually, plants pulled so much carbon out of the atmosphere that the world was pushed into a glacial state in the Early Permian, about 300 million years ago. Not only are fossil plants the source of coal, but most of the world's oil was produced by the bodies of ancient planktonic algae that died and was locked into the rocks. Without plants, there would be no stored energy in the form of fossil fuels on which our society depends.

Over the course of the past 400 million years, plants have been Earth's great thermostat, growing and flourishing to pull excess carbon out of the atmosphere in a greenhouse world and releasing carbon through decay and burning to keep the planet from freezing over completely. The great stores of carbon locked up in the earth's crust as coal (and also as limestone, produced by marine plants and animals) is the switch on that thermostat. Sadly, we are now burning huge amounts of the coal that was stored in the crust for at least 300 million years, and this is pushing the planet back into a greenhouse world, upsetting the natural balance that had been stable for millions of years.

MICROBIAL FOSSILS

The simplest fossils of photosynthetic organisms are produced by sticky mats of microbes, especially blue-green bacteria (cyanobacteria), and by simple algae. Most of their fossils are extremely tiny and can only be seen under a microscope. The only visible fossils they leave are layered structures known as *stromatolites* (fig. 20.2A–B). They look a bit like cabbages

Figure 20.2 ▲

Stromatolites are layered, cabbage-shaped structures formed by sediment trapped by layers of sticky cyanobacterial mats. (*A*) The finely laminated structure of a stromatolite is diagnostic. (*B*) For most of the history of life on Earth, the shallow nearshore waters hosted the only life, mounds and domes of sticky cyanobacterial mats that formed stromatolites. (Illustrations by Mary Persis Williams)

when sliced open, with fine millimeter-scale layers in flat mats, or in domes or even pillars. They are the oldest megascopic fossils on the planet, dating back at least 3.5 billion years, and from then until about 700 million years ago (for almost 3 billion years), they were the only visible life on Earth.

Stromatolites form from sticky mats of bacteria or algae that grow where the light penetrates the sea surface, usually in lagoons and shallow seafloors in the intertidal zone. When the tide is high and they are immersed in water, tiny filaments of bacteria or algae grow up through the sediment washed on top of the older part of the sticky mat, making a new sticky mat on top of the old sediment. Some stromatolites can trap a new layer of sediment every day, although usually they do not grow this fast for very long. Stromatolites are important indicators of ancient environments and are the earliest megascopic record of life.

ALGAL FOSSILS

Algae are the earliest form of plants; they are simple structures that have no hard woody tissues to raise them out of the water. Algae tend to be mostly tiny microscopic cells, but some can make huge structures, such as the submarine forests made of the fronds of kelp. They do not have many of the specialized features found in more advanced plants, so they must always live in water, or near water in a moist habitat.

Most algae do not have hard mineralized tissues, so they do not fossilize. However, some planktonic algae, like diatoms and coccolithophorids, do leave their tiny shells behind. Diatoms may bloom and reproduce and then die in the trillions to make a rock called diatomaceous earth, or diatomite, which is a densely packed mix of pure diatoms. It is mined for water filters, kitty litter, and many other purposes. Coccolithophorid algae bloomed in huge numbers in shallow oceans (especially during the Cretaceous Period) to produce large volumes of limestone made up of their shells, which we know as chalk. The famous White Cliffs of Dover are made of Cretaceous chalk, and there are important chalk beds from the Cretaceous of Belgium and France as well as in the United States in Alabama and Texas and western Kansas.

The most commonly collected megascopic algal fossils are structures known as a *receptaculitids*, nicknamed "sunflower corals" (fig. 20.3A–B). For many years, paleontologists puzzled over these strange structures, which

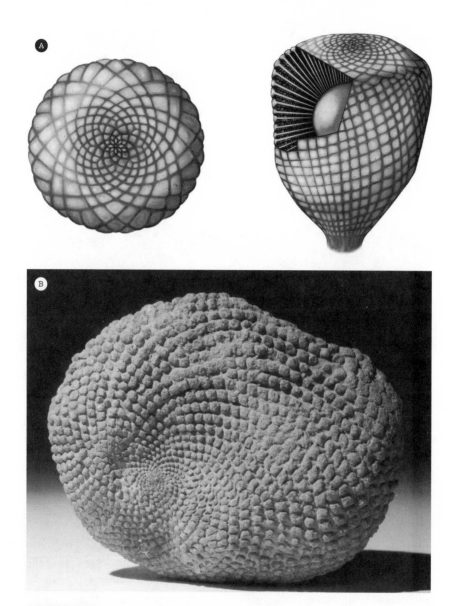

Figure 20.3 ▲

The receptaculitids, or "sunflower corals," are not corals at all but a structure built by a dasycladacean alga. (*A*) Reconstruction of the structure of a complete specimen, showing the spiral patterns on the surfaces. (*B*) A typical specimen, showing the sunflower-like pattern on the surface. ([*A*] Illustration by Mary Persis Williams; [*B*] courtesy of Wikimedia Commons)

look vaguely like a coral head, with a distinctive spiral pattern of cells on top that reminds one of a sunflower head. In older fossil guidebooks, these fossils were often placed with the corals or with the sponges or indicated as a mystery. Recent research, however, has shown that they are the product of a group of living plants known as the dasycladacean algae. Receptaculitids were particularly important as reef builders during the Ordovician, and they are often collected as large complete specimens in Ordovician beds all over the Midwest, from the Cincinnati Arch to Indiana, Tennessee, Wisconsin, and Iowa.

VASCULAR PLANTS: FERNS, CLUB MOSSES, AND HORSETAILS

Algae must live and reproduce entirely in water, and they have no structures that allow them to stick up above the land surface. Neither do mosses, which are among the most primitive of land plants. Mosses cannot grow tall and must also remain moist at all times. A great breakthrough occurred when plants evolved a vascular system of tubes running up long stems to bring water and nutrients from the roots, transport the material produced by photosynthesis up in the leaves, and then move it back through the rest of the plant. Eventually, they evolved stiff woody tissues made of the protein *lignin*, which makes them rigid enough to stand taller and taller. These plants are known as vascular plants, and they are the first plants to grow fully on land and to be able to stand tall above the surface. (Refer to figure 20.1 to review the evolution of land plants.)

The first vascular plants, or *tracheophytes*, are simple fossils such as *Rhynia* or *Cooksonia*, from the Late Silurian and Early Devonian (fig. 20.4A–B). They are not much more than a short vertical stem a few centimeters tall, topped by a reproductive structure at the top, with only minimal branching. They had no leaves, so they must have conducted photosynthesis through the surface of their stems.

One of the first successful groups of vascular plants is the *lycophytes*, represented today by about a thousand species of club mosses and quillworts. They are moisture-loving, low-growing, ground plants and are quite common in certain places, although most nonbotanists do not notice them. One genus, *Lycopodium*, or the "ground pine," is a typical creeping plant in the undergrowth of northern forests (fig. 20.4C). These low-growing

Figure 20.4 ▲

The earliest vascular plants. (*A*) Specimen of *Cooksonia* and (*B*) a reconstruction of the fossil, showing the long stems without leaves topped by the spore-bearing sporangia. (*C*) The living club moss, *Lycopodium* and (*D*) a reconstructions of the giant lycophyte tree *Lepidodendron*, which reached heights of more than 120 feet (36.6 meters). (*E*) Typical lycophyte trunk, showing the diamond-shaped leaf scars on the bark, and (*F*) reconstruction of another common lycophyte, *Sigillaria*, showing the vertical scars up the trunk. (*G*) The living horsetail or scouring rush *Equisetum*, and (*H*) *Annularia*, a fossil horsetail from the Carboniferous. (*I*) Reconstruction of the tall Carboniferous sphenophyte tree *Calamites*. ([*A*] Courtesy of Hans Steuer; [*B*] courtesy of N. Tamura; [*C, E, G–H*] courtesy of Wikimedia Commons; [*D, F, I*] illustrations by Mary Persis Williams)

Figure 20.4 ▲
(continued)

Figure 20.4 ▲
(*continued*)

evergreen plants grow in long strands like ivy and are popular for making Christmas wreaths and garlands. They are more advanced than *Rhynia* and the primitive vascular plants because they are the first plants to have leaves and roots.

Fossil lycophytes first appeared in the Late Silurian and Devonian and were the first vascular plants to rise more than a few feet above the land surface. By the Carboniferous, they grew to be the first large trees on Earth; some were more than 120 feet (36.6 meters) tall with trunks more than 6 feet (2 meters) in diameter (fig. 20.4D–E). They were the dominant plants in the great coal swamps of the Carboniferous Period, and their fossils are easy to collect and recognize in old spoil piles from coal mines. One of the best known is *Lepidodendron*, the "scale tree," which had distinctive dia- mond-shaped scars on the bark of the trunk from old leaves that had fallen off (fig. 20.4D–E). Its leaves were short and blade-like and attached directly to the trunk in a spiral pattern. *Sigillaria* was a slightly shorter lycophyte, with clusters of leaves arranged at the ends of its branches and leaf scars that appear to be arranged in vertical lines up the trunk (fig. 20.4F). After dominating the swampy regions of the Paleozoic, lycophytes declined in the Mesozoic and were replaced by more advanced trees. Most were only shrub-sized or smaller and were found in Mesozoic wetlands. Only the smaller herbs such as *Lycopodium* survive today.

Slightly more advanced are *ferns*, which have a rigid stem and many leaves that branch in pairs from the stem. Ferns first appeared in the Carboniferous Period, and they quickly radiated into a huge diversity of different forms. More than 12,000 species are alive today. Like other prim- itive vascular plants, they do not have seeds but spores that require moist conditions to germinate and grow, so ferns are usually found in damp conditions in or near water. Ferns are often fossilized in swampy deposits around coal seams, and fern fossils are common and easy to collect if you are in the right place. Most ferns today are quite small, although tree ferns still grow in the damper forests of the world. During the Carboniferous, the huge fern *Psaronius* reached 33 feet (10 meters) in height.

A third important group of early vascular plants are the *sphenophytes*, known from the modern "horsetail" or "scouring rush" (genus *Equisetum*). There are only about 20 living species, and most are found in swampy wet- lands today (fig. 20.4G–H). The pioneers called them "scouring rushes" because their stems are full of silica abrasive, and they could be wadded up to make a scouring pad for washing dishes in pioneer days. Sphenophytes have long straight hollow stems separated into segments by joints, and their leaves radiate out from the joints. Most living species are short (barely 3 feet tall or less), but in the Carboniferous coal swamps there were huge trees called *Calamites*, which were 66 feet (20 meters) tall (fig. 20.4I).

GYMNOSPERMS: CONIFERS AND THEIR RELATIVES

All these primitive vascular plants (lycophytes, ferns, sphenophytes) reproduced with spores and required moist conditions and a film of water for the sperm to reach the egg. They were successful in the moist, swampy conditions of the Carboniferous, but they could not survive far from large bodies of water. During the Late Devonian, plants began to evolve a key adaptation that enabled them to breed in drier conditions: the seed. The first seed plants were the progymnosperms, which had a simple naked seed that did not require water for fertilization. They also had true woody tissue in their trunks. In the Late Devonian in places like the Gilboa Forest in the Catskill Mountains of New York, the trunks of huge progymnosperm trees like *Archaeopteris* were over 40 feet (12 meters) tall (fig. 20.5A).

Another group of late Paleozoic plants had even more advanced seeds. They are mislabeled the *seed ferns*, but they are not related to true ferns because they have seeds instead of spores (fig. 20.5B). They were particularly common in the Carboniferous and Permian, especially in drier uplands habitats that were too arid for swamp plants like lycophytes, sphenophytes, or true ferns. The most famous of the seed ferns was a fossil known as *Glossopteris* ("tongue leaf" in Greek), which was found on all the Gondwana continents during the Permian (fig. 20.5C). It was one of the earliest fossils to suggest continental drift. Both of these early seed plants are now extinct.

By the Late Permian and especially in the early Mesozoic, a new group called *gymnosperms* ("naked seeds" in Greek) had appeared (see fig. 20.1). They almost all have their seeds within cones, so many of them are called "conifers." In most cases, distinct male and female cones have different reproductive features. When the male cone releases pollen into the air, the female cones open, and their sticky sap traps the airborne pollen, bringing it to the egg where it is fertilized.

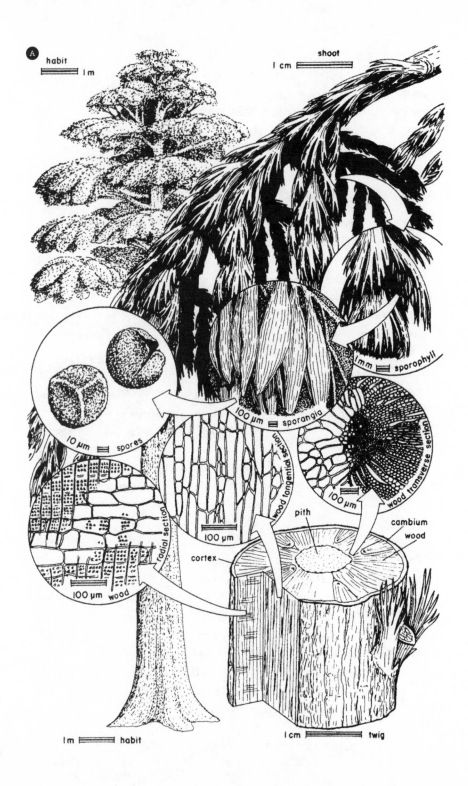

habit

1 m

shoot

1 cm

1 mm   sporophyll

100 μm   sporangia

100 μm

wood transverse section

10 μm   spores

wood tangential section

100 μm

radial section

pith

cambium

wood

cortex

100 μm wood

1 m   habit

1 cm   twig

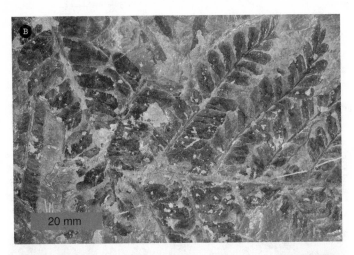

20 mm

Bladafdrukken

Figure 20.5 ◀ ▲

Fossil gymnosperms had primitive seeds, which allowed them to reproduce on dry land. (*A*) Reconstruction of the progymnosperm *Archaeopteris*, showing the details of the trunk, leaves, and roots. (*B*) The fossil seed fern *Neuropteris* from the Carboniferous, and (*C*) the tongue-shaped fossil leaves of *Glossopteris*, a common Permian seed fern on all the Gondwana continents. (*D*) Typical log from the Petrified Forest of Arizona, showing the original tree rings of this relative of the modern *Araucaria*. (*E*) Living *Araucaria*, also known as the Norfolk Island pine, and (*F*) close-up of the distinctive needles of *Araucaria*. (*G*) The distinctive duck-foot-shaped leaves of the maidenhair tree, *Ginkgo biloba*. (*H*) Two modern cycads, with a large round female cone on the left and a tall narrow male cone on the right, and (*I*) the fossilized trunk of a cycad from the Jurassic of South Dakota. (*J*) The archaic conifer *Metasequoia*, known as the "Dawn Redwood"; (*K*) close-up of the seeds of *Metasequoia*; and (*L*) cone of *Metasequoia* from the late Eocene Badger's Nose flora, Warner Range, northeastern California. (*M*) The modern bald cypress *Taxodium*, and (*N*) fossil *Taxodium* branches. ([*A*] Courtesy of G. Retallack; [*B–C, G, I–K, M–N*] courtesy of Wikimedia Commons; [*D–F, H*] photographs by the author; [*L*] courtesy of J. Myers)

Figure 20.5 ▲
(*continued*)

E

Figure 20.5 ▲
(*continued*)

Figure 20.5 ▲
(*continued*)

Figure 20.5 ▲

(*continued*)

Figure 20.5 ▲

(*continued*)

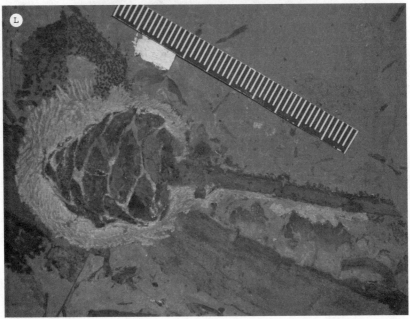

Figure 20.5 ▲
(continued)

Figure 20.5 ▲

(*continued*)

Figure 20.5 ▲
(*continued*)

There are more than 650 living species of gymnosperms, including pines, firs, spruce, and hemlock. The largest living organisms on Earth, the giant sequoia trees (which can be 340 feet [110 meters] tall and weigh many tons), and the oldest living organisms on Earth, the bristlecone pines (which can survive up to 10,000 years), are also gymnosperms. Conifers are very hardy, successful trees whose narrow needles enable them to conserve water and survive in cold conditions where other plants cannot live. They first appeared in the Late Carboniferous, and in the Permian large conifers like *Cordaites* were over 80 feet (25 meters) tall with leaves more than three feet (1 meter) long. Conifers took over nearly all the drier upland habitats in the Triassic and Jurassic. The famous fossil wood of the Petrified Forest National Park in Arizona (fig. 20.5D) comes from the same family as the living *Araucaria* (the Norfolk Island pine), which is now a common tree in cultivated areas, although originally it was found only in and around southern New Zealand and South America (fig. 20.5E–F). Fossil pines were abundant in the Jurassic beds of North America and were often fossilized in Cenozoic plant localities as well.

In addition to *Araucaria* and pine trees, there are a number of other types of gymnosperms. The most familiar is the ginkgo tree, or "maidenhair tree," *Ginkgo biloba*, with its distinctive "duck's foot" shaped leaves (fig. 20.5G). Today it is found in cities and gardens all over the world, and it is also popular as an herbal medicine. Until 1690, when botanists discovered it on the grounds of a Chinese temple and spread it around the world, it lived only in a few remote areas of eastern Asia. Although is it a gymnosperm with simple small cones, it has broad leaves rather than needles. In addition, it is deciduous not evergreen; the leaves turn yellow and are shed abruptly in a few days in the fall, and they grow back in the spring. Many people see ginkgo trees in the cities and in their gardens and do not realize it is a living fossil more closely related to a pine tree than to any other tree.

The other important group of Mesozoic gymnosperms is the *cycads*. These are common in gardens today, where they are sometimes called "sago palms," but they are gymnosperms, not palm trees. Each cycad plant is either male or female and generates a male or female cone when they breed (fig. 20.5H). Cycads have a stumpy trunk with scars of old branches that resemble a pineapple and large palm-like fronds that are very stiff and fibrous. Today they are found mainly in the tropics and subtropics, and they are common in gardens in frost-free parts of the world. But during the Jurassic, they were one of the most common plants of all (fig. 20.5I), and their distinctive stumps (along with a similar-looking plants called cycadeoids) are often found in the dinosaur-bearing formations of the Rocky Mountains, such as the Upper Jurassic Morrison Formation and the Lower Cretaceous Dakota Group.

One of the most common fossil gymnosperms from the Mesozoic were the archaic conifers known as the dawn redwood, or *Metasequoia* (fig. 20.5J–L). Even though it looks superficially like other kinds of conifers, it is very distinctive. It was first known from fossils, but in 1944 it was found alive in the Hubei Province of China, a true living fossil. It has been planted all over the world now, another living fossil (like the ginkgo) that survived in China and then spread to many different regions. Another important Mesozoic conifer is the bald cypress *Taxodium*, which is frequently found in swamps today (fig. 20.5M–N). It is also common in Cretaceous swamp and floodplain deposits where dinosaur fossils are found.

ANGIOSPERMS: FLOWERING PLANTS

All of the vascular plants I have mentioned are less than 5 percent of the diversity of living land plants. Today only one group dominates: the *angiosperms*, or flowering plants. With more than 250,000 living species, they are the majority of plants of most land habitats on Earth. Nearly every plant you see (except conifers and ferns), every plant you find in a nursery or garden, and every plant we use for food is an angiosperm. But angiosperms were a comparatively late development in plant evolution. They arose in the mid-Mesozoic, but by the Late Cretaceous, they had pushed the conifers into the background.

Angiosperms had several adaptations that gave them advantages over other plants. The most important adaptation was the invention of the flower—a highly modified reproductive organ with the ovary in the center and the male reproductive parts on the outer area. The flower co-evolved with Cretaceous insects, such as bees and butterflies, and these insects pollinated the plants. When an insect visits a flower for its nectar bait, the flower transfers pollen to the insect; when the insect goes to the next flower, it fertilizes that plant. This is a much more efficient way to reproduce than the wind-pollinated gymnosperms or the water-dependent pollination of ferns, mosses, and algae. The embryo in the flower also has a food supply, which enables the seed to germinate even in unfavorable conditions. Angiosperms are also capable of very rapid reproduction and growth, so they can breed or grow back in a few days to weeks, whereas gymnosperms require months or years to grow back after being damaged. During the Cretaceous, duck-billed dinosaurs developed highly efficient pavement-like teeth for grinding plants, and their dino-damage may have given fast-recovering angiosperms an advantage over the gymnosperms. Finally, angiosperms have mastered the trick of vegetative reproduction; that is, the plant can also reproduce by sprouting from a cutting of the stem, or in some cases, an entirely new plant can grow from just a part of the plant stem or root system.

Angiosperms not only dominate the plant world today but also dominate nearly all the plant fossil record in the Cretaceous and the Cenozoic.

Many plant fossil localities in the United States produce excellent fossil leaves or petrified wood, from the coal seams of the Cretaceous and Paleocene of Montana and North Dakota to the legendary Eocene leaf beds of Florissant, Colorado, and the Green River Formation in Wyoming, Utah, and Colorado (fig. 20.6A). Fossil leaves have been found in many

Figure 20.6 ▲

Some typical fossil angiosperms. (A) A frond of the palm *Sabalites*, along with fossil herrings called *Knightia*, from the Eocene Green River lake shales of Utah. It is a relative of the modern palmetto tree, genus *Sabal*. (B) A tropical chestnut, or *Sterculia*, leaf from the late Eocene LaPorte flora, northern Sierra Nevada, California; (C) a plane tree leaf of the extinct genus *Macginitea*, from the LaPorte flora; (D) a leaf of *Alnus*, the alder tree, from the Badger's Nose flora, Warner Range, northeastern California; (E) a laurel leaf and some pine litter from the Badger's Nose flora; (F) a magnolia leaf from the Badger's Nose flora; and (G) a leaf of "Oregon grape," or *Mahonia*, from the Badger's Nose flora. ([A] Courtesy of Wikimedia Commons; [B–C] courtesy of D. Erwin and the University of California Museum of Paleontology; [D–G] courtesy of J. Myers)

Figure 20.6 ▲

(*continued*)

Figure 20.6 ▲
(*continued*)

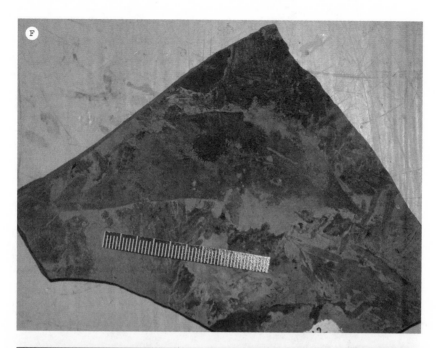

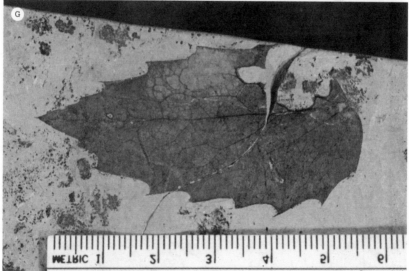

Figure 20.6 ▲
(continued)

other places, including at Eocene and Oligocene plant localities on the Gulf Coast and in Oregon, Washington, and northern California (fig. 20.6B–G). Miocene plant localities are found from the Gulf Coast to the Rocky Mountains to the Pacific Northwest (such as the Ginkgo Petrified Forest State Park in Vantage, Washington, which has petrified wood of 50 species of trees). There are numerous other Pliocene and Pleistocene localities across the United States for collecting fossil leaves or petrified wood. The Calistoga Petrified Forest in Sonoma County, California, has complete logs of giant sequoias that were buried in a volcanic ash eruption 3.4 million years ago. They are the largest known petrified logs in the world.

INDEX